Activities for
ELEMENTARY SCHOOL MATHEMATICS

James W. Stockard, Jr.
Louisiana State University

Vaughn Snyder
Frostburg State University

WAVELAND PRESS, INC.
Prospect Heights, Illinois

For information about this book, write or call:
>Waveland Press, Inc.
>P.O. Box 400
>Prospect Heights, Illinois 60070
>847/634-0081

Copyright © 1997 by Waveland Press, Inc.

ISBN 0-88133-945-8

All rights reserved. No part of this book may be reproduced, stored in a retrieval system, or transmitted in any form or by any means without permission in writing from the publisher.

Printed in the United States of America

7   6   5   4

This book
is dedicated to the authors' families,
to their students, and to their fellow faculty members.

About the authors:

Dr. James W. Stockard, Jr. is an administrator and faculty member in the College of Education at Louisiana State University. He formerly served as professor of education at Auburn University at Montgomery and for many years in the public schools of Shreveport, Louisiana, where he was an elementary teacher, an instructional supervisor, a curriculum director, and an assistant superintendent for curriculum and instruction. Dr. Stockard has written books on elementary social studies, elementary science, and elementary curriculum methods, as well as children's books on science and mathematics. He received his doctorate at Louisiana State University.

Dr. R. Vaughn Snyder is an assistant professor at Frostburg State University in Frostburg, Maryland. He served for five years as an assistant professor at Auburn University at Montgomery. Prior to that he was an elementary teacher specializing in mathematics instruction in Indiana and Ohio. He also conducted math labs for grades 1–6 for Ohio University. Dr. Snyder received his doctorate at Ohio University.

A Note About Games of Chance

While many of the activities in this book require computation, collaboration among pupils, the use of higher-order thinking skills, and the use of problem-solving strategies, "winning" in a particular activity is often merely a coincidence, based on nothing more than luck and chance. Pupils need to know that in games of chance, their skills and abilities have little to do with winning. Pupils should never feel discouraged when they do not win in an activity where winning is purely a matter of happenstance. Winning in such situations is simply a matter of good fortune and luck and has nothing to do with one's skills and abilities.

Table of Contents

PREFACE..xii

CHAPTER ONE
PRENUMBER • NUMBER • NON-NUMBER
- Introduction..1.1
- Number/Numeral Clothespins...1.4
- Patterns...1.6
- Number/Numeral Bingo...1.9
- Number/Numeral/Word...1.12
- Button Sort...1.14
- Odd Man Out...1.16
- Whose Two?...1.18
- Number/Numeral Concentration..1.20
- Walk-on Number Line..1.22
- Names War...1.24

CHAPTER TWO
PLACE VALUE
- Introduction..2.1
- Bean-stick Place Value...2.3
- Tens and Ones...2.5
- Place Value + & –...2.8
- PV Money...2.10
- Roll-a-value..2.13
- Place-value War...2.15
- PV Top This..2.16

- Sum War..2.18
- Place-value "Go Fish"..2.20

CHAPTER THREE
ADDITION • SUBTRACTION

- Introduction..3.1
- + & – Concentration..3.5
- How Many Ways..3.7
- Sub Toss..3.9
- Walk-on Number Line Add/Sub...3.11
- 49 Subtract..3.13
- Addition & Subtraction Tic-tac-toe..3.15
- Dice Addition & Subtraction..3.17
- Marathon Add/Subtract..3.19
- Rollo-sum/Rollo-difference..3.21
- Sum/Difference Penny..3.22
- Overturn/Spill and Add or Subtract...3.24
- Sum/Difference Hold-up..3.26
- Domino Addition/Subtraction...3.28
- Shakedown..3.30
- 100 Adder..3.32
- Sum/Difference War..3.34
- Subtraction is Last...3.36
- Addition/Subtraction Bingo...3.38
- Addition/Subtraction Bingo II...3.40
- Rabbit..3.42
- 2 for 1 Addition/Subtraction..3.44

Activities for Elementary School Mathematics

CHAPTER FOUR
MULTIPLICATION • DIVISION

- Introduction 4.1
- Walk-on Number Line Multiplication 4.5
- Rollo-product 4.7
- Multiplication War 4.9
- Marathon Multiply 4.11
- Multiplication Track 4.13
- Multiplication Match 4.15
- Marathon Division 4.16
- Bobcat! 4.18
- Division Money 4.20
- Divide-a-treat 4.22
- Quotient Bingo 4.24
- Combination Bingo 4.26
- Domino Multiplication 4.28
- Division Rummy 4.30
- Subtractive Division 4.32
- Real Estate Baron 4.34
- Let the Light Shine on Me 4.36
- Divisible Shuffle 4.38
- Cross Off 4.40

CHAPTER FIVE
FRACTIONS: COMMON • DECIMAL

- Introduction 5.1
- Fraction Introduction 5.3
- Fraction Fold 5.4
- Bingo Fractions 5.6

- Fractional Match..5.9
- Equivalent Bingo..5.11
- Decimal/Fraction/Percent Bingo.....................................5.13
- Decimal/Fraction Concentration....................................5.16
- Decimal Tic-tac-toe..5.18
- Concentration Toss...5.20

CHAPTER SIX
GEOMETRY
- Introduction..6.1
- Geo-board Shapes..6.3
- Geo-card Game...6.6
- Shape-o...6.9
- What Shape?...6.11
- Perimeter..6.13
- Area...6.15
- Everyday Geometry...6.17
- Geo-sort..6.19
- Triangle Sort..6.21
- Battleship...6.23
- *PI*, Anyone?...6.25

CHAPTER SEVEN
ESTIMATION
- Introduction..7.1
- Outfit Estimate..7.3
- Class Estimate...7.5
- Water, Water Everywhere..7.7
- Friday Estimate...7.9

- Length Estimate 7:11
- Guesstimate 7.13
- Blinking Eyes 7.15
- Right Price Estimate 7.17
- Estimate/Measure It 7.19
- Oh, How Long, How Wide, How High 7.21

CHAPTER EIGHT
CALCULATIONS

- Introduction 8.1
- Calculator Baseball 8.3
- Inverse Operations 8.5
- Facts Review 8.7
- Calculator + ... 8.8
- Calculator – 8.9
- Wipe-out 8.10
- Target–50 .... 8.12
- Your Order, Please 8.14

CHAPTER NINE
MISCELLANEOUS MATHEMATICAL CONCEPTS

- Introduction 9.1
- Days Are Money 9.3
- Bingo Times 9.5
- Bar Graph 9.7
- I'm For Two 9.10
- Names 9.12
- Large Numbers 9.13

- Integer Golf..9.15
- Sports Card Statistics..9.17
- Probability..9.19
- Spinner Probability...9.21
- Midterm Test Statistics...9.23

CHAPTER TEN
PROBLEM SOLVING
- Introduction..10.1
- Score 33...10.3
- Zoo...10.5
- Price Check..10.7
- Balanced Answer...10.9
- Barges or Trucks?...10.11
- Shoppers..10.13
- Stool Maker..10.15
- Total 9..10.17
- Colorado Mountains...10.19
- Today's News...10.22

APPENDIX
1. Children's Books Helpful in Teaching Mathematics —
   An Annotated Bibliography...11.1
   - Number Sense and Numeration-Grades K-3..11.1
   - Number Sense and Numeration-Grades 4-6..11.2
   - Whole Numbers-Grades K-3...11.3
   - Whole Numbers-Grades 4-6...11.4
   - Geometry and Spatial Sense-Grades K-3..11.4
   - Geometry and Spatial Sense-Grades 4-6..11.5

Activities for Elementary School Mathematics

- Measurement-Grades K-3..11.6
- Measurement-Grades 4-6..11.7
- Statistics and Probability-Grades K-3..11.8
- Statistics and Probability-Grades 4-6..11.8
- Fractions and Decimals-Grades K-3...11.9
- Fractions and Decimals-Grades 4-6...11.10
- Patterns and Relationships-Grades K-3...11.11
- Patterns and Relationships-Grades 4-6...11.12

2. Standards for School Mathematics..12.1
   - Using the Standards...12.33
   - The National Council of Teachers of Mathematics....................12.35
3. Mathematics Resources on the Internet..13.1
4. Constructivism..14.1
5. Multiple Intelligences..14.2
6. Reproducibles..15.1
   - Numerals..15.1
   - Numerals Spelled Out..15.2
   - Numeral Pictures..15.3
   - Order Form..15.4
   - Large Numerals...15.5
   - One-inch Numbered Squares..15.9
   - 7 x 7 Game Board..15.10
   - Blank Bingo Card..15.11
   - Regular Bingo Card..15.12
   - Boat Game Grid...15.13
   - Geometric Shapes..15.14
   - Triangles...15.15
   - Tic-tac-toe Board...15.16
   - Chart..15.17
7. References...16.1

PREFACE

STATE OF THE ART TEACHING IN ELEMENTARY SCHOOL MATHEMATICS

"State of the art" is a goal that every school mathematics program in the United States would say it strives for. It is a goal owed our children and it is attainable. State of the art depends upon curriculum reform. In the current national curriculum reform movement, initiated by the National Council of Teachers of Mathematics and supported at all levels by those involved with mathematics education, it is particularly exciting to be engaged in transforming mathematics teaching and learning to reach the state of the art.

Major curriculum reform is not new in the field of school mathematics. The last such reform was the "new math" of the late 1950s and 1960s which emphasized the unifying mathematical concepts of logic and set theory. For a variety of reasons, the new math did not receive widespread acceptance. Specifically, it did not pay close attention to how students learn and what they are capable of learning at different ages. It also did not address what teachers know about mathematics and pedagogy or how they can best enhance their own knowledge.

The new math was followed by the "back to basics" movement which emphasized rote memorization and the learning of paper-and-pencil algorithms. The current reform grew out of the inability of the back to basics movement to address key issues, such as:
- Neglect of higher order thinking and problem-solving skills;
- Disquieting findings about American students in recent international studies on mathematics achievement, despite the return to basics;
- Changing mathematical skills needed in the work force;
- New research findings on teaching and learning mathematics; and
- The mushrooming of inexpensive calculators and computers.

In an effort to systematically address these issues, the National Council of Teachers of Mathematics contributed standards which have been endorsed by groups representing the mathematics community from kindergarten through graduate school, as well as by many other groups interested in mathematics education. The NCTM standards are presented in their entirety in the Appendix, beginning on page 12.1.

Also in the Appendix, starting on page 11.1, there is an annotated bibliography of children's books which are helpful in the teaching of mathematics. These books were chosen because of their proclivity for integrating mathematics instruction through the book's story line.

Reproducibles, which can be used on a copy machine, are included in the Appendix, beginning on page 15.1.

The activities in this book should help teachers enrich the mathematical experiences of all children and reach toward fulfillment of the NCTM *Standards*.

Activities for Elementary School Mathematics

CHAPTER ONE
PRENUMBER • NUMBER • NON-NUMBER

Prenumber concepts are assumed to be covered in the kindergarten, but it cannot be assumed that all first graders have mastered these concepts.

Classification is the sorting of objects by some characteristic or attribute. Attribute blocks are a useful classification tool. They can be sorted by shape, size, color, or thickness and can include squares, circles, triangles, rectangles, and hexagons. Teachers can make their own sets and select shapes with which their pupils are familiar from class discussions.

Comparison of two sets is also a prenumber concept. When introducing pupils to the concept of comparing sets, the teacher should use two sets that have a great difference in the amounts in each set.

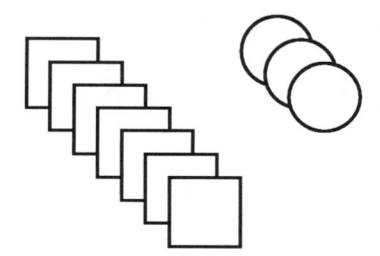

Which set has more items in it? Which set has the fewest items?

Teachers should ask pupils to select sets with the most items and sets with the least items, so pupils realize that sets can be selected for either concept. Thus, pupils look at the sets both ways: more and less.

One-to-one correspondence occurs when pupils match sets of items.

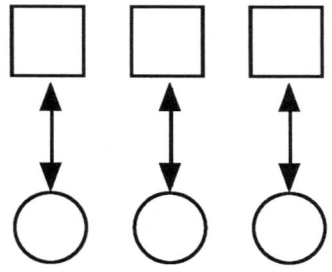

This set contains the same number of squares as it does circles.
Pupils match one circle with one square to do a one-to-one comparison.

Another prenumber concept is *ordering* (or *seriation*), the placing of sets in order from smallest to largest. The pupil begins by finding the set which contains the fewest items and then selects the set with the largest number of items, or vice versa. If, for example, there are three sets altogether, the pupil decides where the third set should be placed (between the smallest and the largest sets).

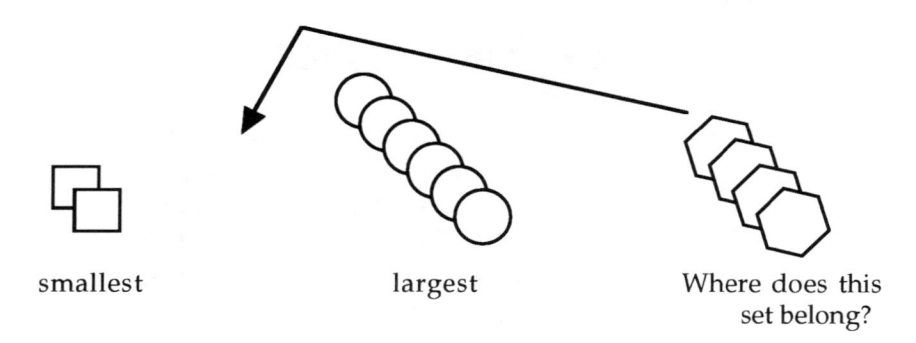

smallest largest Where does this set belong?

Rote counting is considered a prenumber concept and involves learning the number names in order. This can be accomplished by a variety of activities and songs, such as "One, Two, Buckle My Shoe."

Number concepts include such items as number recognition, numeral recognition, number/numeral matching, rational counting, and addition/ subtraction readiness.

Number recognition is the ability to identify how many items are in a set of objects.

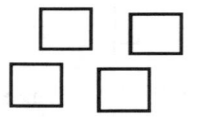

How many squares are there in this set?

Rational counting, simply put, is assigning a name to an item in the set (one, two, three). Rational counting is sometimes called *touch counting*.

Numeral recognition involves identifying a given numeral. Children adept at numeral recognition can correctly respond to such questions as, "Turn the TV to channel four, please," and "Find the seven on the calendar."

Number/numeral matching requires pupils to match the number of items in a set with a corresponding numeral. It helps pupils understand that number is an idea of quantity (how much, how many, etc.) and that a numeral is the name given to that quantity. A numeral is a name for a number. An understanding of the relationship between number and numeral is among the most important concepts in mathematics. This matching process is learned after both number and numeral concepts have been mastered by the pupil.

How many squares in the set? 5, 4 or 6?

Activities for Elementary School Mathematics   Page 1.3

Addition readiness addresses the union of disjoint sets of items. The + and = signs can be introduced after the concept involved is understood. *Subtraction readiness* addresses the separation of sets.

Non-number concepts are geometric shape recognition, spatial relationships, recognition of measurement instruments, and pattern recognition. Initially, *geometric shape recognition* usually covers the basic shapes of square, circle, rectangle, and triangle.

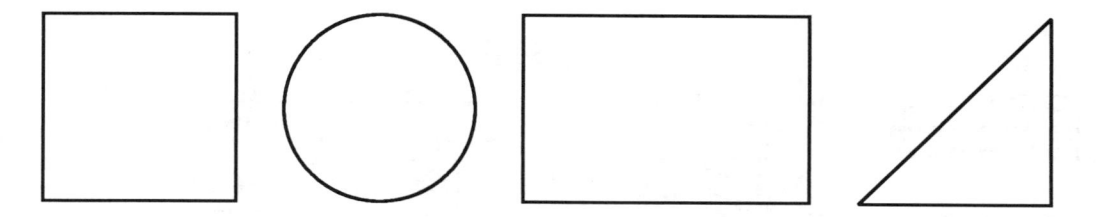

Spatial relationships include such ideas as above, beside, between, over, and under.

Put the book *above* the globe.   Put the book *between* the computers.

Knowing that clocks or calendars are used when discussing time and that a ruler is used to measure length are examples of *recognition of measurement instruments*. *Pattern recognition* involves reproducing patterns or identifying what would be next in a given set. The activities in this chapter address many of these topics.

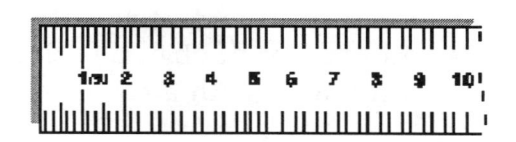

Page 1.4   Activities for Elementary School Mathematics

NUMBER/NUMERAL CLOTHESPINS

Topic: NUMBER/NUMERAL WORD MATCHING

Grade Level: K-1   **Activity Time:** 1 class period

Materials Needed:
1. Ten 3 x 5 index cards
2. Twenty clothespins:

CLOTHESPINS

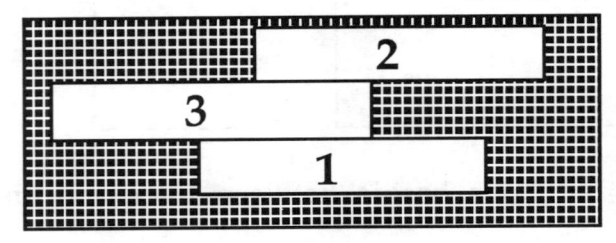

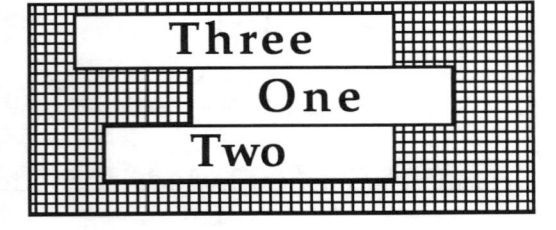

One set of 10 with the numerals 1-10 on them   One set of 10 with the words one-ten on them

3. A bag large enough to hold the clothespins
4. A flat work surface, such as a desk or table
5. Pictures of objects ranging between one and ten items

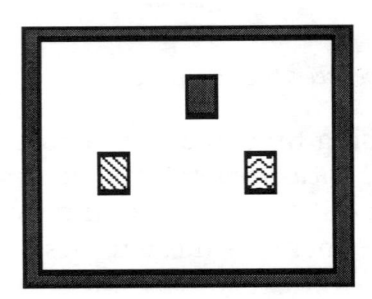

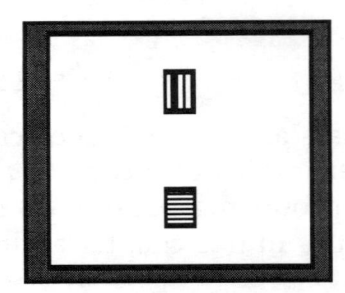

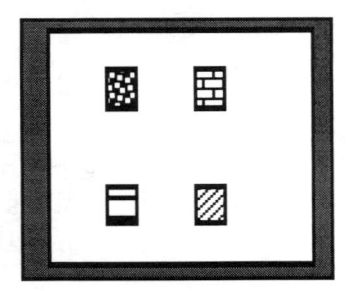

Objectives:
As a result of this activity, the learner will match:
- an amount of objects with a corresponding numeral.
- an amount of objects with a corresponding word name.
- a numeral with a corresponding word name.

In preparation for the activity, paste the pictures onto the 10 index cards (consider laminating them to make them more durable). Before class, mix the picture cards, place them face up on the work surface, and place the clothespins in the bag. For a larger class, use several duplicate sets of clothespins, cards, and bags, and divide the class into two or three groups.

Activities for Elementary School Mathematics   Page 1.5

Introduction:
1. Explain to the class that each pupil will select a clothespin from the bag without looking into the bag.
2. The pupil then attempts to clip the clothespin to the appropriate card.

Major Instructional Sequence:
1. Emphasize that each card may be a match for more than one clothespin. For example, a picture card with four items may be matched by a clothespin displaying the *numeral 4 and* by a clothespin displaying the *word* four.

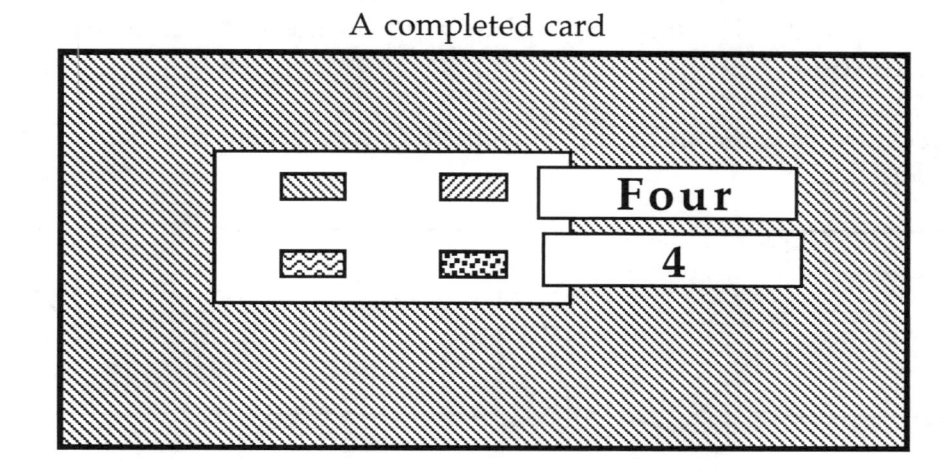

A completed card

2. Each pupil takes a turn selecting a clothespin and attaching it to the appropriate card. If a pupil cannot make the match, classmates may assist in the matching process.

Closure or Evaluation:
The activity continues until all of the clothespins have been correctly placed.

Variation:
Clothespins marked with zero/0 could be included in the set (along with a blank index card) to familiarize pupils with the concept of zero.

Page 1.6   Activities for Elementary School Mathematics

PATTERNS

Topic: NON-NUMBER CONCEPT OF PATTERNING

Grade Level: K-1   **Activity Time:** 1 class period

Materials Needed:
1. A large, flat surface, such as a table
2. A chalkboard
3. Pattern blocks, attribute blocks, or geometric shapes, enough for a small group of pupils (The amount will depend on the number of pupils involved in the lesson.)

NOTE: *Blocks are easy to make and may include any geometric shape that has been covered in class.*

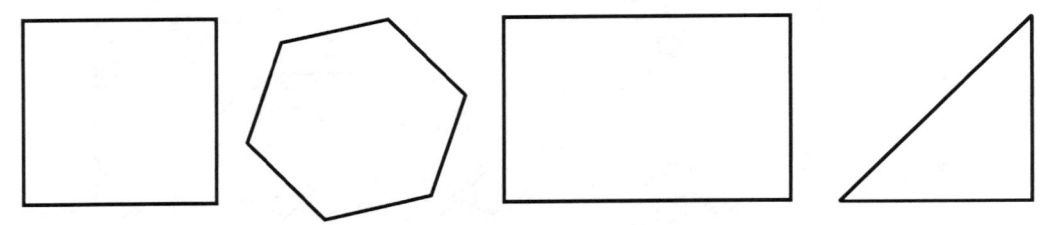

Objectives:
As a result of this activity, the learner will:
- copy a pattern made by the teacher.
- extend a pattern (that is, show what comes next in a certain pattern which has already been established).

Introduction:
1. Explain the meaning of the word "pattern" to the class in a manner that young children will understand. As an example, lay out a simple pattern, such as the one illustrated below, on the table.
   This pattern uses pattern blocks. 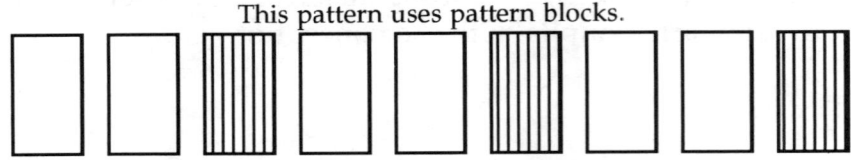
2. Draw the pupils' attention to the fact that the combination of "plain-plain-striped" happens over and over. Explain that this "over-and-over" occurrence is a *pattern*. Tell the pupils that the *pattern* in this group of blocks is plain block, plain block, striped block.
3. Mix the blocks in random order, and ask a pupil or pupils to make the same pattern you first created.

Activities for Elementary School Mathematics   Page 1.7

Major Instructional Sequence:
1. Continue with simple patterns that can be replicated by the pupils without much difficulty.
2. After several simple patterns have been successfully copied, set up a more complex pattern.

This pattern uses attribute blocks or flat, geometric shapes.

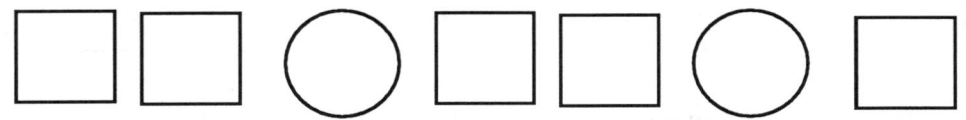

☐ ☐ ○ ☐ ☐ ○ ☐

3. Instruct pupils to copy the pattern and decide what the next two or three blocks or shapes in the pattern would be by placing them correctly on the table.
4. After pupils successfully complete several of the more complex patterns that combine shapes in different series, follow up with even more difficult patterns using blocks that are not only shaped differently but also have different designs on their surfaces.

Shapes such as those in the illustration below are appropriate for hands-on class work. They can be cut out from cardboard, and their patterns may be painted on. The chalkboard is an ideal medium for explaining the process used to create this pattern. Draw the outlines of the shapes, point to them in a rhythmic manner that emphasizes the sequence of shapes, and then draw the correctly chosen design within each.

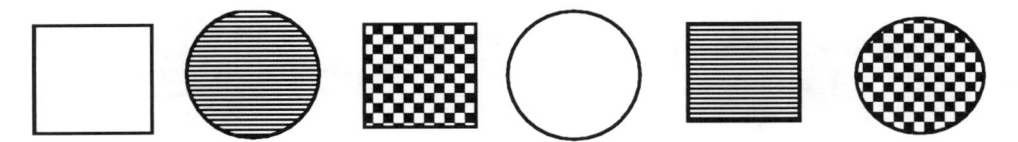

☐ ⊜ ▦ ○ ☰ ◉

5. Ask what would be next in the pattern as established. Prompt the pupils with questions of the following nature:
   - Do you notice a pattern in the order in which these shapes appear? (square/circle, square/circle)
   - Do you notice a pattern in the order in which designs *inside* these shapes appear? (plain/striped/checkered, plain/striped/checkered)
   Combine the answers to these two questions in a manner that enables the pupils to understand the overall pattern and allows them to choose the next item in the pattern.
   - If we decided that the next shape in the pattern will be a square and the next design will be plain, then what will come next in this pattern? (a plain square)

☐

Variation:

If pupils have difficulty grasping the method by which the next item in a pattern is chosen, it may help to visually separate the items into subsets.

To determine the shape: This grouping would encourage pupils to choose the next TWO items in the pattern (another subset of square/circle).

To determine the design: This grouping would encourage pupils to choose the next THREE items in the pattern (another subset of plain /striped/ checkered).

Closure or Evaluation:

Ask pupils to volunteer to come up to the chalkboard, create a pattern, and explain the nature of the pattern to the class. Continue this activity until pupils seem familiar with creating and understanding patterns.

Activities for Elementary School Mathematics

Page 1.9

NUMBER/NUMERAL BINGO

Topic: NUMBER/NUMERAL/WORD MATCHING

Grade Level: K-1

Activity Time: 1 class period

Materials Needed:
1. BINGO-type game cards that have five rows across by five columns down, with a number/set, numeral/symbol, or number/word written in each box (The range of numbers/numerals/words can be as small as 1 through 10 or as large as 1 to 100, depending on class abilities.)

NOTE: *The B-I-N-G-O heading for each column is not needed on this type of card.*

2. 3 x 5 index cards ("calling cards") with a number, numeral, or word on each card, identical to those on the game cards
3. BINGO chips (enough to cover the pupils' game cards)

	one		twenty	5
16	six		fifteen	19
eight		22	8	
20		9	ten	15
	five	6		twelve

An example of a number/numeral BINGO card

Objectives:
As a result of this activity, the learner will match:
- an orally-called number with a corresponding game-card *amount/set.*

Page 1.10   Activities for Elementary School Mathematics

- an orally-called number with a corresponding game-card *numeral/symbol*.
- an orally-called number with a corresponding game-card *number/word*.

In preparation for this activity, use the index cards to create BINGO calling cards, selecting the concepts that have been used in class. (This game makes a good review lesson.) From the calling cards, duplicate the same concepts when creating the BINGO playing cards. Take care that each number you include in your calling cards is reflected in fair proportion on the squares of the BINGO cards.

Introduction:
1. Distribute game cards and chips to the class. Explain to the pupils that the chips are to be used for covering the squares on the game cards as the numbers are called.
2. Act as the caller for the first few games, making sure to mix the numeral cards well before each game starts.

Major Instructional Sequence:
1. Turn the first calling card face up and call out the first numeral to the class. If it is 6, for example, the pupils will look for the numeral 6, the word six, or a set of six items within the squares on their cards.

	one		twenty	5
16	six		fifteen	19
eight		22	8	
20		9	ten	15
	five	6		twelve

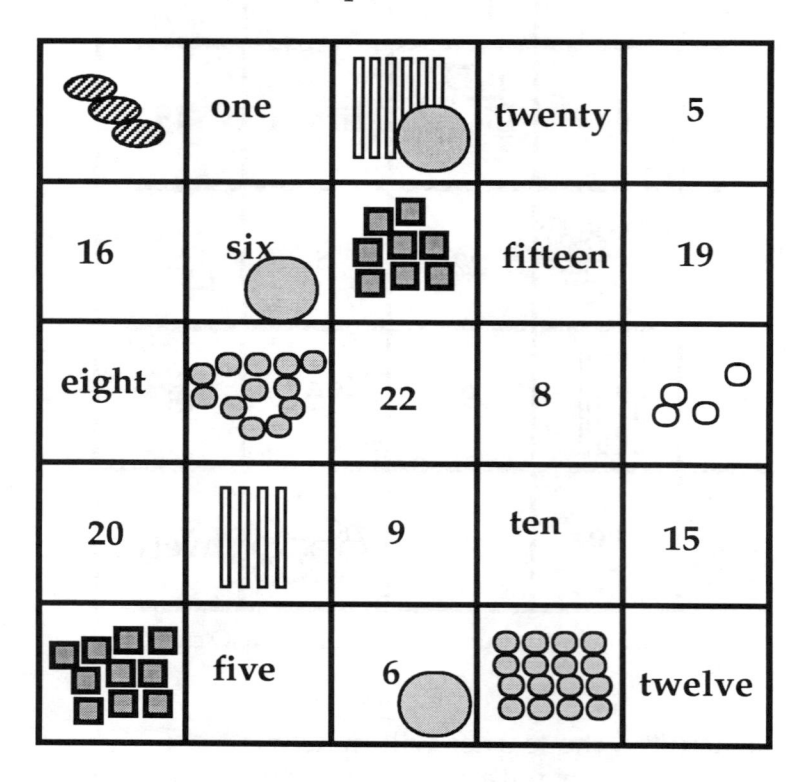

Sample BINGO card with the numeral 6, the word six, and a set of 6 covered

2. Call the second numeral from your stack of cards.
3. Pupils will continue to cover the numbers, numerals, or words with chips as they find the called numbers on their cards.

Closure or Evaluation:
1. The game continues until someone covers five in a row, column, or diagonally on his or her card, at which time he or she shouts, BINGO!
2. Ask the pupil to read what is covered on the winning card in order to verify the BINGO. After a BINGO has been verified, the game cards are cleared and a new round begins.

Variation:
Each round could continue until a second- or third-place BINGO has occurred. The winner of each game may take a turn at being the caller.

NOTE: *The teacher should circulate among the pupils, paying particular attention to children who are continually unsuccessful. Such children can be grouped later for aggressive reteaching of the skills involved, a part of which might be a similar BINGO game which has been especially structured by the teacher to provide success.*

Page 1.12   Activities for Elementary School Mathematics

NUMBER/NUMERAL/WORD

Topic: NUMBER/NUMERAL/WORD MATCHING

Grade Level: K-2   **Activity Time:** 1 class period

Materials Needed:
1. Ten 6 x 8 cards with a numeral ranging from 1 through 10 on each card
2. Ten 6 x 8 cards with pictures of items ranging in amount from 1 through 10 on each card
3. Ten 6 x 8 cards with a word ranging from one through ten on each card (If working with a larger class, duplicate cards may be necessary, or increase the range to between 1 and 20.)

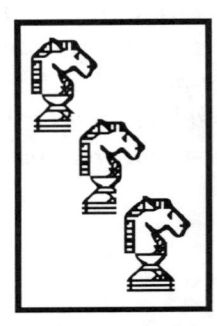

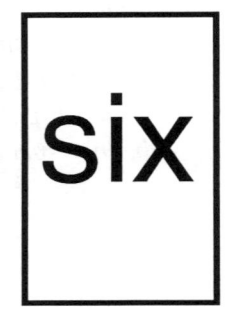

3	♞♞♞	six
numeral card	picture card	word card

Objectives:
As a result of this activity, the learner will:
- match number to numeral.
- match numeral to word.
- identify number, numeral, and word.

Introduction:
1. Choose and shuffle two of the sets of cards to distribute to the class: for example, retain the numeral cards but distribute the picture cards and word cards to the pupils.
2. Tell the class that you will hold up a card from the set you retained, and anyone whose card matches your card is to raise his or her hand.

Major Instructional Sequence:
1. As you hold up a card, ask the children whose hands are raised to hold up their cards and take turns orally identifying them: for example, "I have three stars on my card," "My card has the word THREE on it," or "The numeral 3 is on my card."

Activities for Elementary School Mathematics　　　　　　　　　　　　Page 1.13

2. Check to make sure that each response is correct. (Make a mental note or keep a written, pupil-by-pupil record of responses in order to determine which pupils are having difficulty with the matching process.)
3. The activity continues until all children have had a chance to identify their cards and view all of the other matches.

Closure or Evaluation:
When all matches have been made, collect, resort, and redistribute the cards. Keep a different set of cards from the set you previously retained to hold up and be matched by the pupils.

five

FIVE

5

Page 1.14   Activities for Elementary School Mathematics

BUTTON SORT

Topic: CLASSIFICATION SKILLS

Grade Level: K-2   **Activity Time:** 1 class period

Materials Needed:
1. A large receptacle containing buttons of different shapes and sizes (Parents and grandparents are good sources for buttons. Ask each pupil to bring in 20 or 30 buttons from home.)
2. A work surface, such as a table top, on which to place the buttons

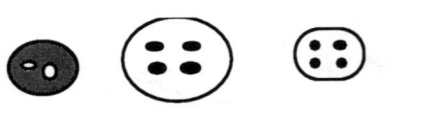

Objectives:
As a result of this activity, the learner will:
- classify objects by attribute (their characteristics; features that make objects alike and different).

Introduction:
1. Ask the class to place all their buttons into the large container.
2. Separate pupils into groups of two or three and give each group a handful of buttons scooped from the large container (about 50-60 buttons).
3. Tell the pupils that they are going to sort the buttons into groups according to an attribute. Explain that an attribute is a feature that makes the buttons alike and different. Show that some possible attributes of buttons are size, color, and number of holes.

Major Instructional Sequence:
1. First ask the pupils to put the buttons into groups according to color. Some colors may not be easy to place and some buttons may be multicolored, so a discussion will be needed to select groups in which to put the odd-colored buttons. A category of "others" would be acceptable for those buttons that do not fit into any of the basic color groups.
2. Monitor the pupils, making sure that everyone is active in placing buttons into the groups.
3. After buttons have been classified by color, ask pupils to sort them into subgroups according to size. Small, medium, and large are appropriate size groups, but be aware that some buttons may be a source of discussion in order for the pupils to reach a decision about their size.

Activities for Elementary School Mathematics Page 1.15

Closure or Evaluation:
>Make sure the attributes are understood by all pupils and discuss any buttons which cause problems.

>Lead the class in a discussion about the attributes of other things they may have in common, like shoes, book bags, pets, and the like.

Variation:
>Play money in the denomination of pennies, nickels, dimes, quarters, and half-dollars may also be used to improve pupils' classification skills.

Page 1.16 Activities for Elementary School Mathematics

ODD MAN OUT

Topic: NUMBER/NUMERAL/WORD MATCHING

Grade Level: 1-2   **Activity Time:** 1 class period

Materials Needed:
1. Eleven 6 x 8 index cards, with a *numeral* ranging from 0 through 10 on each card

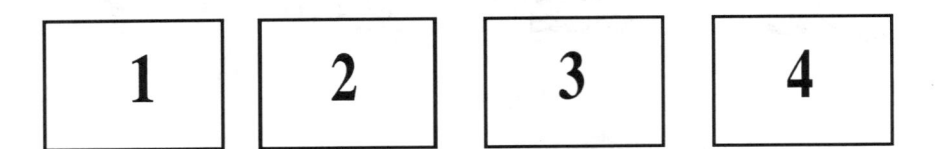

   | 1 | 2 | 3 | 4 |

2. Eleven 6 x 8 index cards, with a *word* ranging from zero through ten on each card

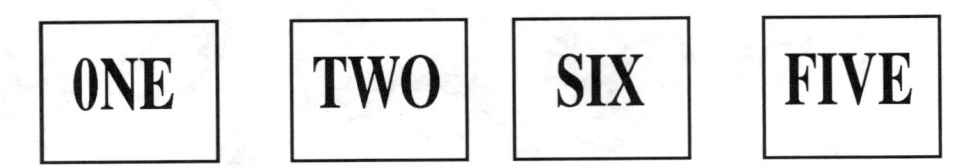

   | ONE | TWO | SIX | FIVE |

3. Eleven 6 x 8 index cards, with a *picture* ranging in amount from 0 to 10 objects on each card (The zero card would be blank.)

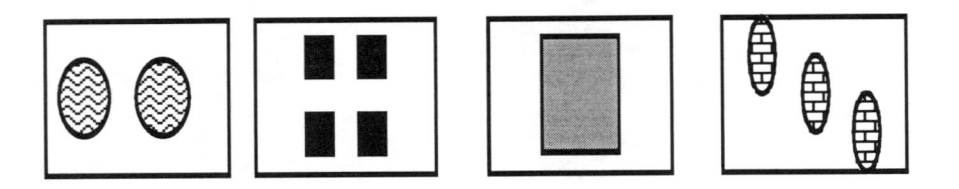

4. Paper and pencil for each pupil in class

Objectives:
As a result of this activity, the learner will develop:
- number/numeral matching skills.
- numeral/word matching skills.
- number/word matching skills.

Introduction:
1. Before class, presort the cards into sets of four, with three cards belonging to a

Activities for Elementary School Mathematics   Page 1.17

set and the fourth card *not* belonging for example, the numeral 4, the word four, a picture of four items, and the word three.
2. Tell the class that they will look for a member of a set that does not belong with other set members.

Major Instructional Sequence:
1. Place a set of four cards in the chalk tray for the class to view from their desks.

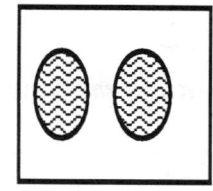

| 2 | ONE | ◉ ◉ | TWO |

2. Instruct the pupils to think about the set and copy onto their papers the card that does not belong. (Remind pupils to write large enough so that you can see their responses from the front of the room.)
3. On your signal, ask all pupils to hold up their choice for you to view from the front of the class.
4. Then show the class which card did not belong in the set and explain why it did not belong (for those who may have missed the correct answer). Encourage discussion, taking care not to embarrass any pupils who missed the correct response.

Closure or Evaluation:
The game continues until all desired numbers have been covered by the class.

Variation:
The set size could be increased, or the pupils could look for *two* items that do not belong with the others.

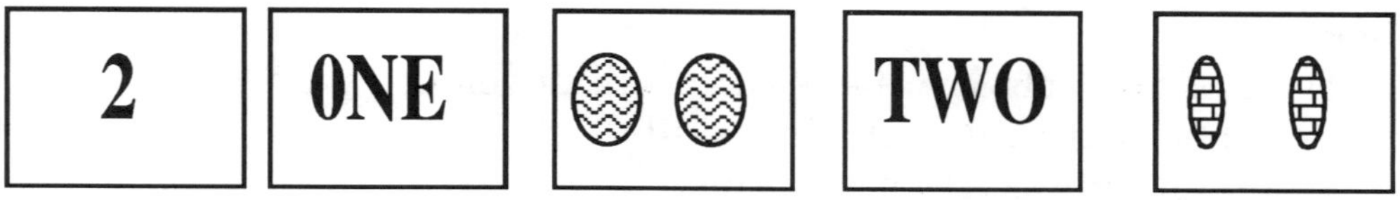

| 2 | ONE | ◉ ◉ | TWO | ◉ ◉ |

| 2 | ONE | ◉ ◉ | TWO | ◉ ◉ | ▓ |

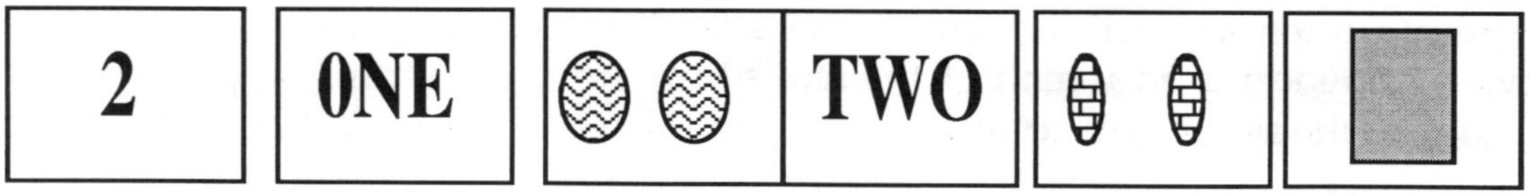

Page 1.18   Activities for Elementary School Mathematics

WHOSE TWO?

Topic: NUMBER/NUMERAL/WORD MATCHING

Grade Level: 1-2   **Activity Time:** 1 class period

Materials Needed:
1. Thirty-three 6 x 8 index cards (if working with a larger class, duplicate cards may be necessary):

   Eleven cards with a *numeral* ranging from 0 through 10 written on each of them

   | 1 | 2 | 3 | 4 |

   Eleven cards with a *word* ranging from 0 through 10 written on each of them

   | ONE | TWO | SIX | FIVE |

   Eleven cards with groups of objects ranging in number from 1 through 10 *pictured* on each of them (The eleventh card is blank, representing zero.)

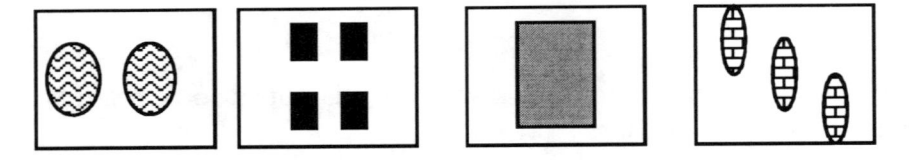

Objectives:
   As a result of this activity, the learner will develop:
   - number/numeral matching skills.
   - numeral/word matching skills.
   - number/word matching skills.

In preparation for this activity, create the sets of index cards, making more or less of one type, depending on which of the skills you are concentrating (number/numeral matching, perhaps, or any combination of the three concepts listed in the objectives).

Activities for Elementary School Mathematics   Page 1.19

Introduction:
1. Distribute the index cards to the class.
2. Tell the class that you will call selected numbers aloud and that they should look for those numbers on their cards.

Major Instructional Sequence:
1. Call out a number (such as two).
2. Instruct each pupil who has the numeral 2, the word two, or a set of two objects on his or her card to stand and show the class what is on the card.
3. Call out a new number and repeat the process.

Closure or Evaluation:
The game continues until all desired numbers have been identified.

🐸🐸🐸🐸   four 4

🐯🐯   two 2

🤠🤠🤠
🤠🤠🤠   six 6

🚚   one 1

Page 1.20   Activities for Elementary School Mathematics

NUMBER/NUMERAL CONCENTRATION

Topic: NUMBER/NUMERAL/WORD MATCHING

Grade Level: 1-2   **Activity Time:** 1 class period

Materials Needed:
1. Paper and pencil for each pupil in class
2. A flat playing surface, such as a large desk or table, for each group of 2 to 5 pupils
3. Forty 3 x 5 cards for each group of pupils:

| 1 | 2 | 3 | 4 |

10 cards with a number between one and ten written on each of them in *numerical* form

| ONE | TWO | SIX | FIVE |

10 cards with a number between one and ten written on each of them in *word* form

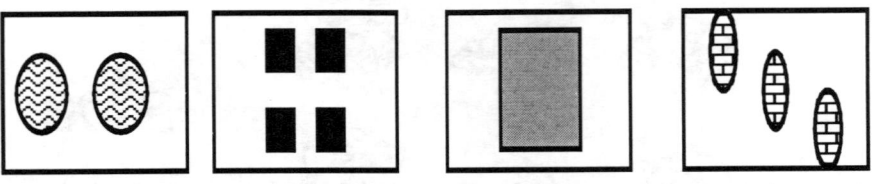

Two sets of 10 cards with a number of objects between one and ten *pictured* on each of them

Objectives:
As a result of this activity, the learner will develop:
- number/numeral-matching skills.
- numeral/word-matching skills.
- number/word-matching skills.

In preparation for this activity, set up a CONCENTRATION game board with eight rows of five cards placed face down for each group of pupils.

Activities for Elementary School Mathematics

Introduction:
Tell the class that they are going to play MATH CONCENTRATION.

Major Instructional Sequence:
1. Explain to the pupils the process of searching for matches on the board: Each pupil in the group will take a turn looking for matches by turning over any two cards at a time. If the cards match (for example, 1/one, or two/2), the pupil who finds that match keeps the cards and play moves to the next player. If no match is made, the cards are returned to their original face-down position on the playing surface and the next pupil takes a turn. (Emphasize the importance of remembering the location of the cards in order to make a match.) Begin the game.
2. Circulate among the groups of pupils, checking for understanding and giving assistance as necessary.
3. The game continues until all of the cards have been matched.

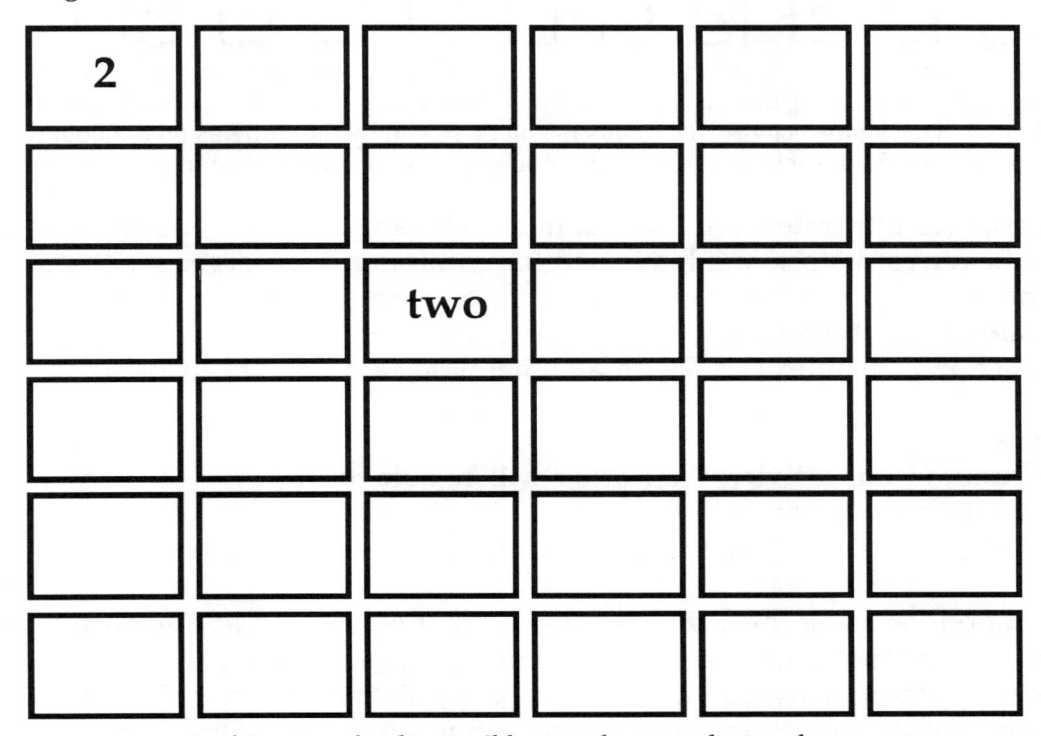

In this example, the pupil has made a match, **2** and **two**.

Closure or Evaluation:
The winner from each group is the pupil with the most cards at the end of the game.

Variation:
By using *only* the 10 numeral cards *or* the 10 word cards with *one* set of 10 picture cards, a game board of five rows by four columns of cards may be assembled for younger pupils.

WALK-ON NUMBER LINE

Topic: RATIONAL COUNTING

Grade Level: 1-3

Activity Time: 1 class period

Materials Needed:
1. A walk-on number line

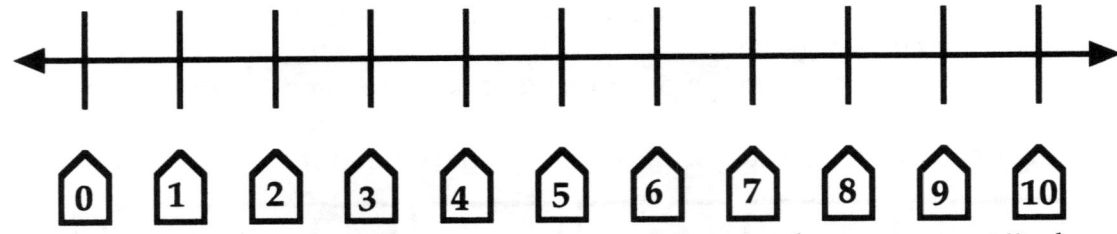

The walk-on number line can be made of heavy plastic, such as an old shower curtain. Affix the number line to the floor with masking tape, using black tape for the center line and cross lines, spaced so pupils can walk on it. During the activity, the numbers will be placed by the students on the floor next to the number line, with their arrow-like sides pointing to each appropriate intersecting line.

2. Cards (as illustrated above) with numerals ranging from 0 through 10, made of heavy card stock and laminated for endurance (For larger groups of children, make two sets of cards, one set for each side of the number line. The children can then walk on each side of the center line.)
3. A set of 3 x 5 index cards with a numeral on each card, from 1 - 10

Objectives:
As a result of this activity, the learner will be able to:
- rationally count to ten.
- determine numbers that are greater than, less than or between.

In preparation for using this activity, consider that as a first-grade activity the pupils will work on rational counting on the number line, progressing to addition and perhaps subtraction. Any activity that can be conducted on a regular number line can be accomplished on the walk-on number line. As an activity for the older children, an introduction to numbers which are greater than or less than others can also be undertaken on the number line by encouraging interactive dialogue with questions such as: "I have 3. Who has a number greater than mine? Who has a number less than mine?"

Introduction:
1. For first graders:
   - Ask pupils to take turns walking down the number line silently.

Activities for Elementary School Mathematics

- Ask pupils to walk down the number line a second time and rationally count as they take each step.
2. Form pupils into groups of ten.
3. Distribute the numeral cards to the first group.

Major Instructional Sequence:
1. Ask, "Who has the number 3?" That pupil walks out to the third cross-line on the number line as indicated by the number card on the floor. Continue asking questions such as, "Who has 7?" and encourage pupils to take their places on the number line. Proceed in this manner until all pupils in the group are on the number line, then have them give you their numeral cards and stand away from the number line.
2. Shuffle the numeral cards, distribute them to the next group, and repeat the process.
3. Recycle the groups and continue until each child has had several different numbers.
4. For older children, ask:
   - "Who has a number greater than 7?"
   - "Who has a number less than 7?"
   - "Who has a number between 3 and 7?"
   - Continue in this fashion until each pupil has had several positive experiences.

Closure or Evaluation:
Give all pupils the opportunity to answer questions.

> **NOTE:** *See chapters 3 and 4 for addition/subtraction and multiplication/division activities on the number line.*

NAMES WAR

Topic: OTHER NAMES FOR A NUMBER

Grade Level: 2-4   **Activity Time:** 1 class period

Materials Needed:
1. A pair of dice
2. Paper and pencil for each pupil
3. A stopwatch or one-minute timer

Objectives:
As a result of this activity, the learner will:
- determine as many ways as possible to represent a number.

Introduction:
1. Provide each pupil with paper and pencil. Roll the dice. (If conducting this activity with a large group, you may want to draw the "roll" on the chalkboard as in the illustration below, so that the whole class is able to see it clearly.)
2. Instruct the pupils to write on their papers as many "names" as possible for the number rolled. Depending on the grade level, fractions, decimals, or division may be used. Any "name" utilizing a mathematical process to arrive at the rolled number is permissible.

If three is rolled, encourage pupils to think of as many mathematical representations for three as they can. Some possible answers for this roll are 4-1, 2+1, 7-4, 3 x 1, 3/1, or 21÷7.

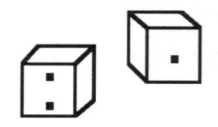

Major Instructional Sequence:
1. Time the pupils (thirty to sixty seconds per roll).
2. Circulate among the pupils, checking for understanding and giving assistance as necessary.

Closure or Evaluation:
1. One point is awarded for each correct response.
2. A time limit may determine the length of the game, or a total-point goal may decide the winner.

Activities for Elementary School Mathematics

CHAPTER TWO
PLACE VALUE

A numeral has a value depending on its location. For example, in the numeral 326, the "2" is in the tens column and has a value of twenty, or two tens. To assist children in the understanding of place value, a place-value mat (such as the one illustrated below) is a valuable learning tool. The place-value mat can be very helpful when used in conjunction with bean sticks and beans, base ten blocks, or poker chip trading.

Bean sticks and beans are very useful for helping pupils to understand place value. To make a bean stick, simply glue beans to popsicle sticks. It is best to use beans of uniform size and color. Bundles of ten Popsicle sticks or drinking straws can also be used for place-value manipulatives.

PLACE-VALUE MAT		
Hundreds	Tens	Ones

On a place-value mat the columns are labeled as hundreds, tens, and ones.

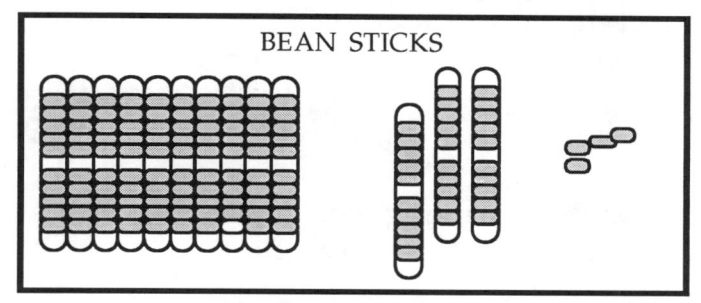

BEAN STICKS

The number represented here is 134; one bean raft containing 100 beans, three ten-sticks representing 30, and four individual beans.

Base ten blocks use the same principle as the bean sticks and beans, with the units cubes representing the ones, the rods representing the tens, and the flats representing the hundreds. Regrouping by addition and subtraction is accomplished by trading in a "tens" rod for ten unit cubes, which would be placed in the ones column.

Poker-chip trading utilizes chips of different colors to represent different place values. Chip trading is more abstract than other place-value manipulatives, since the objects on the mat are the same size but have an assigned value by color. Chip trading should be used only after the concept of place value has been developed with blocks or sticks. When working with pupils, one might assign the white chips a value of one hundred each, the blue chips a value of ten each, and the red chips a value of one each.

The number represented below is 246 because there are two white chips (with a value of 100) representing two hundred, four blue chips (with a value of ten) representing forty, and six red chips (with a value of one).

PLACE-VALUE MAT		
Hundreds	Tens	Ones
OO	●● ●●	◯◯◯ ◯◯◯

As shown below, bundle sticks or straws in groups of ten, then put ten groups of ten together and place a rubber band around the group to represent a group of one hundred. Below are two groups of one hundred, four groups of ten, and four individual straws; 244.

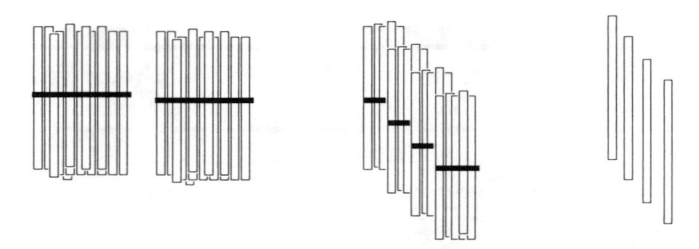

This chapter employs these learning tools in activities that should stimulate children's interest in the concept of place value.

Activities for Elementary School Mathematics

Page 2.3

BEAN-STICK PLACE VALUE

Topic: PLACE VALUE

Grade Level: 1-2

Activity Time: 1 class period

Materials Needed:
1. Beans, bean sticks, and bean-stick rafts (if children are working with three-digit numbers) enough for each pupil or group to have 1 of the hundreds rafts, 20 of the 10-sticks, and at least 30 individual beans.

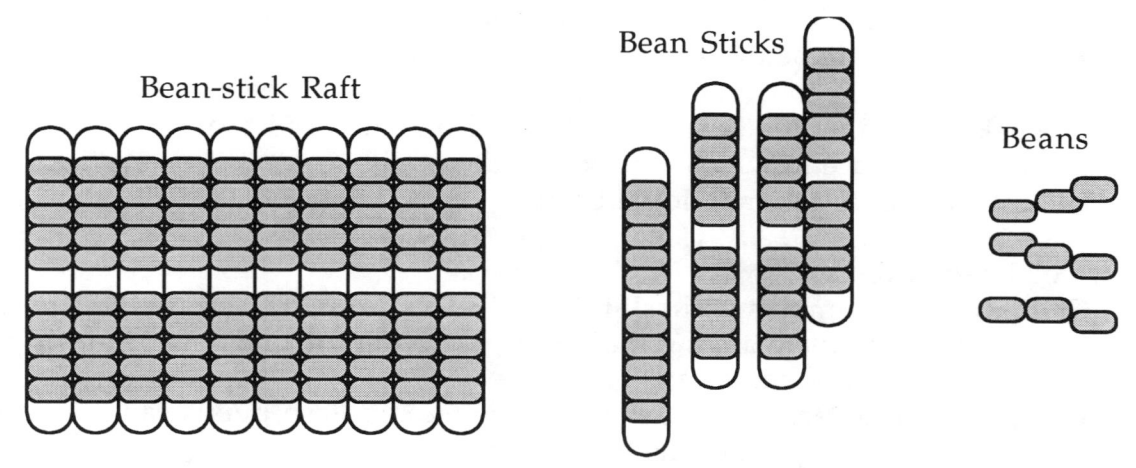

Bean-stick Raft   Bean Sticks   Beans

The number represented here is 149; one raft representing 100, four ten-bean sticks representing 40, and nine individual beans representing 9 units, or ones.

NOTE: *If you are working with a large class, pupils could be put into groups of three.*

2. Place-value mats for each pupil or group
3. Paper and a pencil for each pupil or group

Objectives:
As a result of this activity, the learner will:
- have a better understanding of numbers in the base ten system.
- write a base ten numeral for the number represented on the place-value mat.

Introduction:
1. Distribute *only* the individual beans to the pupils.
2. Challenge the pupils to a race with you. Write a numeral on the chalkboard, such as 24, and at your signal race the pupils (or groups) to count that number of individual beans.
3. Pick up two of the ten-sticks and four individual beans and declare yourself the winner.

Major Instructional Sequence:
1. Encourage discussion about why you were the winner of the race. Hold up the ten-sticks and explain that you used two groups of ten to represent twenty. Draw a simple place-value mat on the chalkboard, illustrating two ten-sticks in the tens column and four individual beans in the ones column. Using the illustration, explain how twenty can be represented by two groups of ten.
2. Distribute the bean sticks (and bean-stick rafts, if the class is working with hundreds) and place-value mats.
3. Review the material with the pupils:
   - Ask the pupils (or groups) to count the number of beans on one of their bean sticks to verify that each stick contains ten beans.
   - If hundreds are being studied, ask pupils to guess how many beans are on the raft. Then explain that the raft contains ten bean sticks, with each stick containing ten beans, which equal one hundred beans.
4. Have another race in which you permit the pupils to use the bean sticks and you use only the individual beans.

Closure or Evaluation:
1. Call out other numbers and ask the pupils to represent them on their place-value mats. Circulate among the pupils, checking for understanding, and giving assistance as necessary.
2. If appropriate, select some numbers in the hundreds so pupils can use and understand the purpose of the bean-stick raft.

> **NOTE:** *Bean sticks and rafts can be used in addition, subtraction, multiplication, and division. Base ten blocks (if the class has access to them) can be used as well, as can popsicle sticks or straws.*

Activities for Elementary School Mathematics   Page 2.5

TENS AND ONES

Topic: PLACE VALUE

Grade Level: 1-2   **Activity Time:** 1 class period

Materials Needed:

1. Place-value mats with ones, tens, and hundreds for each pupil (They can be created on the computer and printed or reproduced on regular typing or copier paper.)

PLACE-VALUE MAT		
Hundreds	Tens	Ones

2. Red, white, and blue poker chips (twenty of each color for each pupil)

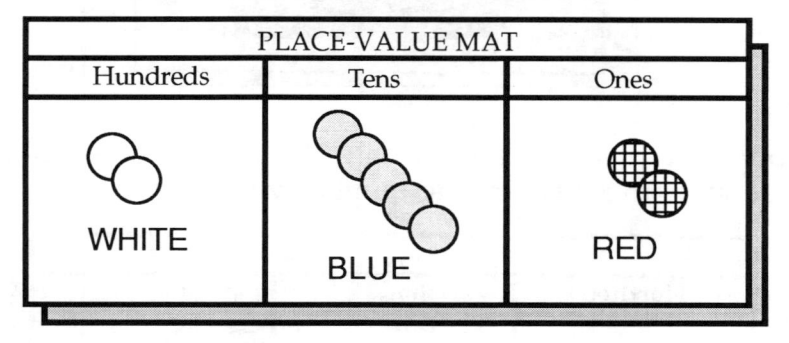

PLACE-VALUE MAT		
Hundreds	Tens	Ones
WHITE	BLUE	RED

○ This chip has been assigned the value of 100.

◯ Tens have been assigned to chips of this color.

⊕ Ones have been assigned to chips of this color.

The number 252 is represented on the place value mat using the concept of chip trading.

3. Paper and pencil for each pupil

> **NOTE:** *Base ten blocks or bean sticks could also be used in this exercise.*

Page 2.6   Activities for Elementary School Mathematics

Objectives:
As a result of this activity, the learner will:
- become familiar with tens and ones in the place-value system.
- be able to regroup numbers on a place-value mat in the tens and ones columns.

Introduction:
1. Provide each pupil with a place-value mat.
2. Distribute 19 "ones" chips and 8 "tens" chips to each pupil.
3. Divide chalkboard into three sections representing hundreds, tens, and ones.

Major Instructional Sequence:
1. On the chalkboard, write a *number* between 1 and 8 in the "tens" column and a *number* between 10 and 19 in the "ones" column — for example, a 3 in the "tens" column and a 16 in the "ones" column.

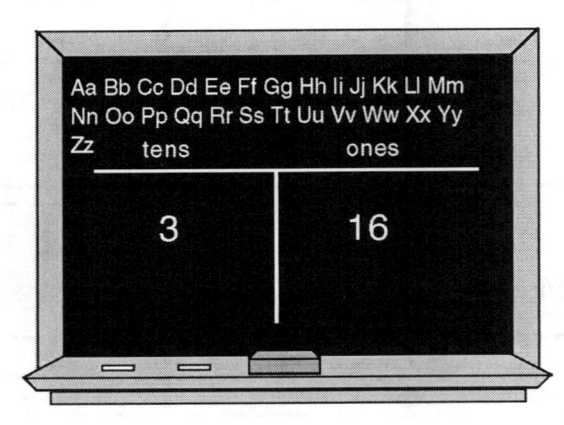

2. Ask the pupils to place the appropriate "tens" and "ones" chips on their place-value mats to represent the number you wrote.

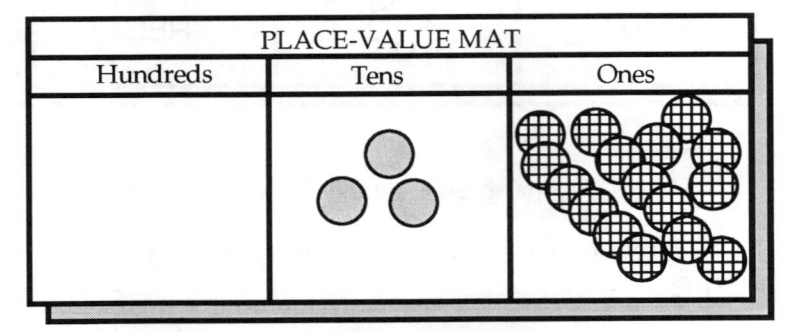

This place-value mat shows 3 tens and 16 ones.

3. When they are ready, instruct the pupils to regroup the chips from the "ones" column to the "tens" column.

Activities for Elementary School Mathematics

PLACE-VALUE MAT		
Hundreds	Tens	Ones
	●●●●● (4 circles)	● ● ● ● ● ● (6 grid circles)

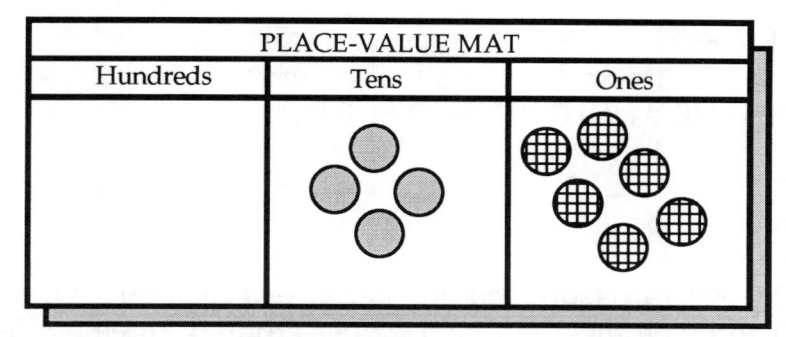

Regrouped, the number would be four tens and six ones, or forty-six.

4. Circulate among the pupils, checking for understanding, and giving assistance as necessary.

   NOTE: *If this activity is used early in the school year, you may choose to use numbers which do not require regrouping until pupils become more comfortable with the concept of place value.*

Closure or Evaluation:
1. Explain the regrouping process, using the chalkboard to illustrate if necessary. Write the resulting new *number* on the chalkboard (for example, 46), and ask the class to say the number out loud.
2. After the class understands the basic concept of place value and can easily regroup numbers from the "ones" to the "tens" column, select numbers containing "tens" that would require regrouping in the "hundreds" column. Continue the activity, selecting other numbers for the class to regroup until the pupils seem comfortable with the concept of place value.

PLACE VALUE + & -

Topic: PLACE VALUE ADDITION AND SUBTRACTION

Grade Level: 2-3   **Activity Time:** 1 class period

Materials Needed:
1. Sixty 3 x 5 index cards, divided into decks: 6 of each card, with a numeral between 0 and 9 on them — one deck per group of three pupils (The amount of cards in each deck will vary, depending on the size of the class, but each group should have a minimum of 12 cards.)

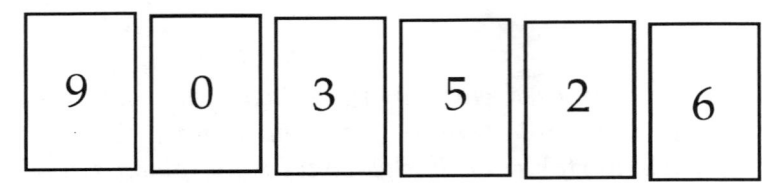

2. Pencil and paper for each pupil
3. Calculators (one for each group of three pupils)

Objectives:
As a result of this activity, the learner will:
- create and read three-digit numbers.
- add and subtract two numbers, each of which is three digits.
- verify addition or subtraction answers with the use of a calculator.

Prior to this activity, make sure that the decks of cards are divided up fairly (for example, ensure that one deck does not contain all the nines).

Introduction:
1. Divide pupils into groups of three. For each group, choose a distributor who will be responsible for passing out the cards and checking addition and subtraction with the calculator.
2. Place a deck of cards in a stack next to each group's distributor.
3. Instruct the distributor to pass out three cards to each of the other two players in his or her group.

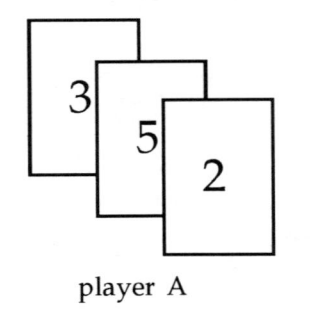

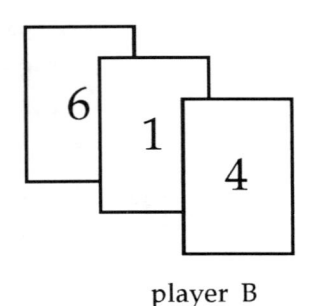

        player A   player B

Activities for Elementary School Mathematics   Page 2.9

Major Instructional Sequence:
1. Instruct pupils to create a three-digit number with the cards they have received. Players may rearrange their cards in any order to make a new number at any time *before* they record their numbers on their papers.
2. After the first numbers are recorded on the pupils' papers, the distributor will pass out three more numbers to each player, who will then create another new three-digit number.
3. The distributor indicates whether addition or subtraction is to be accomplished in order to obtain a third number. (Players are not told in advance which operation will be selected for each round of the activity.)
    - If the operation is subtraction, the distributor instructs players to subtract the smaller number from the larger one.
4. The winner of the round is the pupil from each group with the largest number after the addition/subtraction is accomplished. The answers of both pupils are checked by the distributor, with the aid of a calculator. Winners receive one point, and scores are kept by the distributor.
5. Rotate to the right to choose a new distributor for each round — the new distributor should be the player to the right of the pupil who was the distributor for the first round. The players will return their cards to the deck after each round, the new distributors should shuffle their decks well, and play continues.

Closure or Evaluation:
The winner of the game is the player from each group who has the most points after ten rounds. If winning players from different groups have tie scores, act as distributor for a tie breaker. Deal three more cards to each tied player, asking them to create the largest number possible. The winner is the person who creates the largest number.

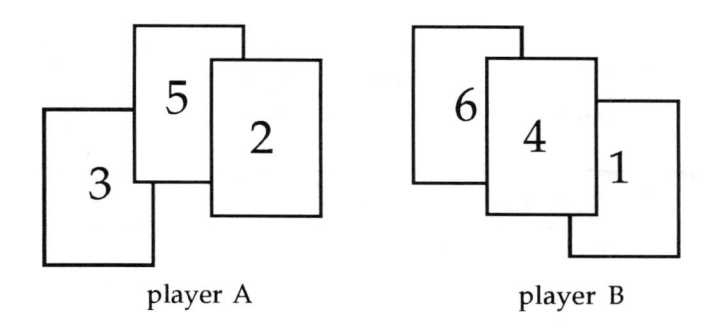

player A   player B

With these two sets of cards, if pupils made the largest number possible and recorded it, player B would have been the tie-breaking winner.

PV MONEY

Topic: PLACE VALUE AND MONEY

Grade Level: 2-3   **Activity Time:** 1 class period

Materials Needed:
1. Spinners with ten places on them, marked 0-9 (one for each group of four pupils)
2. Place-value mats with "ones" and "tens" columns on them (one for each pupil)

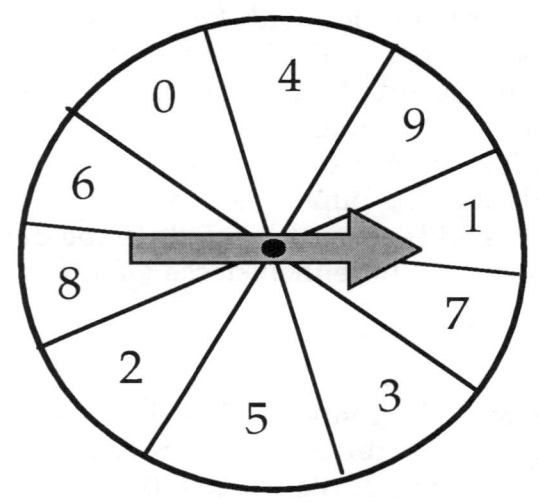

3. Paper and pencil for each pupil
4. Play money representing pennies, dimes, and dollars

NOTE: *Dollar bills may be included if you choose a higher amount of money as the goal in this activity. It is an effective way for young children to learn about money in conjunction with place value.*

For older pupils, place-value mats should include a "hundreds" column.

Activities for Elementary School Mathematics   Page 2.11

Objectives:
   As a result of this activity, the learner will:
   - become familiar with place value.
   - become familiar with values of pennies and dimes.
   - learn to regroup in the base ten system.

Introduction:
   1. Divide pupils into groups of four.
   2. Distribute place-value mats, paper, pencils, and play money representing a one-dollar bill, nine dimes, and nine pennies to each pupil.
   3. Distribute one spinner to each group.
   4. Explain that the object of the game is to be the first pupil in a group to obtain coins in an amount equal to one dollar or more.
   5. Pupils from each group spin to see who gets highest spin, and that person starts the activity.

Major Instructional Sequence:
   1. Instruct the pupils to place the play money in a pile in the middle of their playing surface.
   2. The first pupil from each group spins, selects coins that are equal to the value of his or her spin, and place the coins on his or her place-value mat in the appropriate columns.

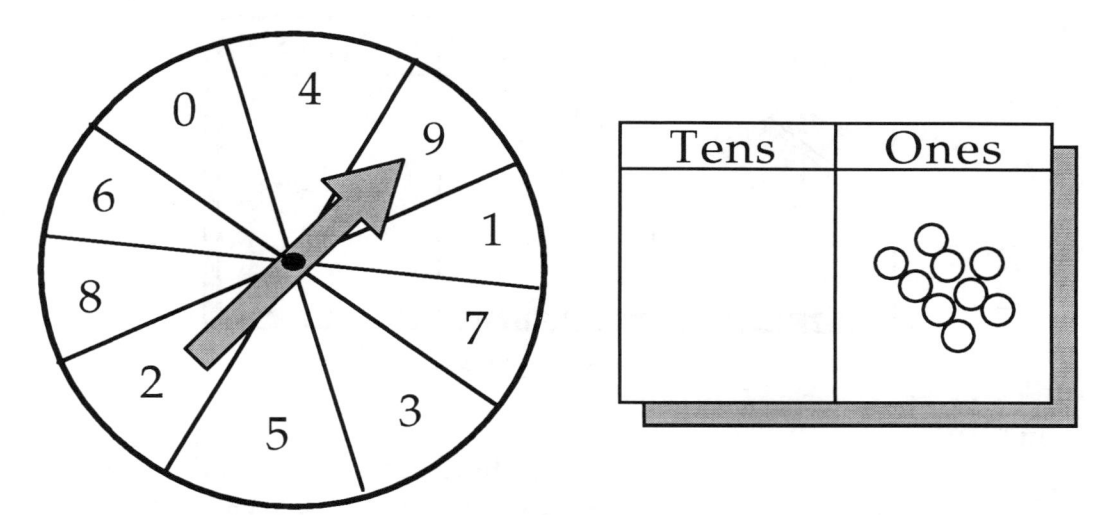

   This player spun a nine and placed nine pennies in the "ones" column.

   3. Play continues to the right and eventually returns to first player in each group. These players spin again and place an amount of money equal to their second spins on their mats. If the coins on a pupil's mat exceed ten in the "ones" column (pennies), he or she may trade in ten of the pennies for one dime and place the dime in the "tens" column on the place-value mat. (Paper and pencil may be used to verify players' answers.)

4. Circulate among the groups, checking for understanding, and giving assistance when appropriate.

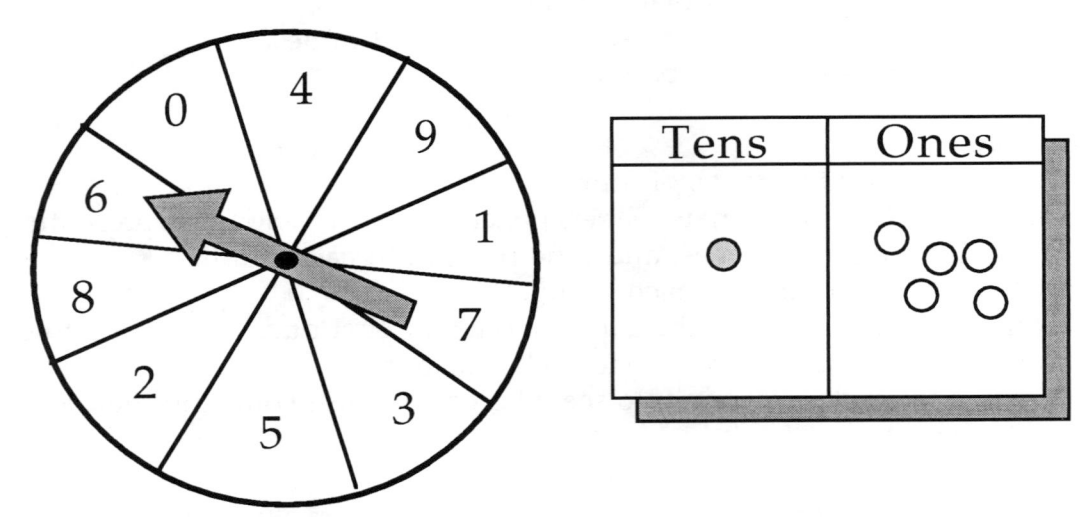

In this instance the player spun a six and regrouped as needed.

Closure or Evaluation:
The first player in each group who obtains coins equal to one dollar or more is the winner.

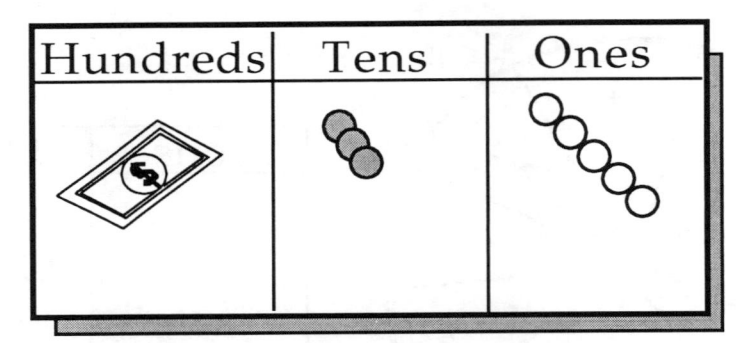

Here, 135 pennies is represented by $1.35.

Variation:
For older children, not only one-dollar bills but also ten-dollar bills and hundred-dollar bills (such as those used in a Monopoly game) may be used to play to a goal of up to one thousand dollars (or whatever limit is appropriate for the age and expertise of the players).

Activities for Elementary School Mathematics

ROLL-A-VALUE

Topic: PLACE VALUE

Grade Level: 2-3   **Activity Time:** 1 class period

Materials Needed:
1. A pair of dice for each group of four pupils *(One die in the pair must be a different color from the other.)*
2. A place-value mat for each pupil
3. A record sheet for each pupil (see illustration on the following page)
4. Base ten blocks (or bean sticks) for place values of hundreds, tens and, ones for each group of four pupils
5. A calculator for each group of four pupils

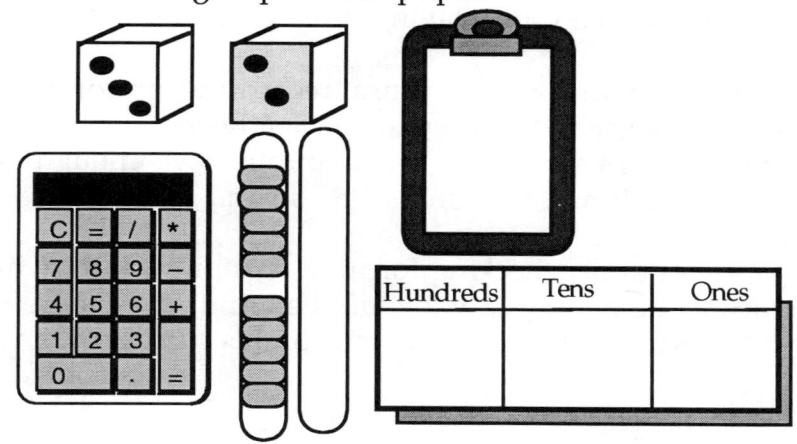

Objectives:
As a result of this activity, the learner will:
- have a better understanding of place value.
- use a calculator for addition.

Introduction:
1. Separate pupils into groups of four. Each pupil in his or her group rolls one die to see who rolls the highest numeral (the highest roll will start the activity).
2. Ask each group to choose one color of die to use as the "tens" indicator and the other to determine the "ones" values. (The color selection will stay the same for all rounds and should be recorded on their record sheets for future reference.)

Major Instructional Sequence:
1. Instruct the first pupil within each group to roll the pair of dice and record on his or her paper the results of the roll according to the colors for the tens and ones the group selected.

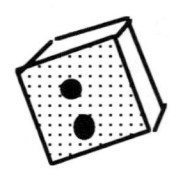

Page 2.14   Activities for Elementary School Mathematics

On these dice, if the pupils had selected
the clear die to represent the tens, the number represented
would be thirty-two. If the clear die had been selected to represent the ones,
the number would be twenty-three.

2. The first player from each group records the results of his or her initial roll on a record sheet and places the equivalent base ten rods and units cubes (or bean sticks and individual beans) on the place-value mat.
3. The player to right of the first player rolls next, following the same procedure, with the players in each group taking turns until play returns to the first players.
4. The first players take their second turn, record the appropriate information on their record sheets, and again place an equivalent amount of beans or blocks on their place-value mats. If any pupils have enough items to regroup in either column, they should do so and check their work, using the calculator.
5. Circulate among pupils, checking for understanding, and giving assistance.
6. When a place-value mat has a value of one hundred or more on it, the player must use a "hundreds" flat (or bean raft) on his or her place-value mat.

Closure or Evaluation:
1. Play continues until a player from each group reaches a goal of one thousand.
2. You may set a time limit or select a smaller number than one thousand for the goal if you desire to shorten playing time.

After three rolls of 65, 46, and 14,
a pupil's record sheet and place-value mat would look like this.

PLACE-VALUE MAT		
Hundreds	Tens	Ones

clear die	other
1. 6	5
2. 4	6
11	1
3. 1	4
4.	

The pupil's next move would be to combine the amount of the new roll
with the total of the previous rolls to get a new number on the place-value mat
and on the record sheet, checking the work with the calculator.

Activities for Elementary School Mathematics

PLACE-VALUE WAR

Topic: PLACE VALUE

Grade Level: 2-4   **Activity Time:** 1 class period

Materials Needed:
1. 3 x 5 index card "decks" of 10 cards with a numeral ranging from 0 to 9 on each — one "deck" of 10 cards for each pair of pupils)

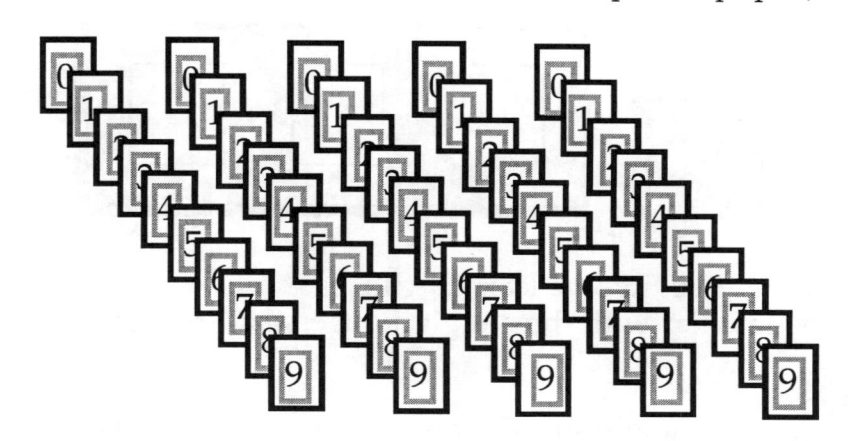

Objectives:
As a result of this activity, the learner will:
- recognize which number is largest.

Introduction:
1. Divide the pupils into pairs.
2. Shuffle each deck of cards and place one deck in the middle of each pair of pupils.
3. Each pupil draws one card to determine who starts the activity (the highest cardholder in each pair starts, and the game proceeds, alternating turns).

Major Instructional Sequence:
1. Instruct the first pupil in each pair to take the top two cards from the stack and arrange them to create the largest two-digit number possible.
2. The next player from each pair does the same, and the round continues.
3. Circulate, check for understanding, and give assistance when appropriate.
4. The player with the highest number wins all of the cards from the round.

Closure or Evaluation:
Play continues until all of the cards from the deck have been drawn. The winner is the player with the most cards.

Page 2.16   Activities for Elementary School Mathematics

PV TOP THIS

Topic: PLACE VALUE

Grade Level: 2-4   **Activity Time:** 1 class period

Materials Needed:
1. Dice with numerals on them (one die for each group of three or four pupils)
2. Pencil and game card (a laminated game card is best) for each pupil

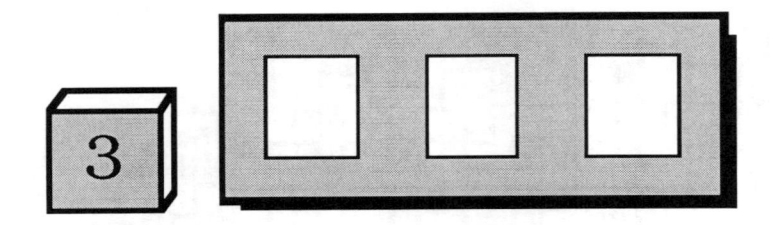

Objectives:
As a result of this activity, the learner will:
- become familiar with three-digit numbers and recognize which are larger.

Introduction:
1. Separate pupils into groups of three or four and distribute one die to each group.
2. Provide each pupil with a game card like the one below.
3. Explain to pupils that they will each get four turns at rolling the die. The objective of the game is to create the largest number possible after four rolls, using three of the four numbers rolled.

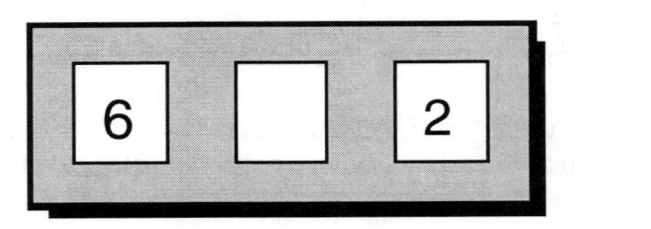

passed number

Major Instructional Sequence:
1. Instruct the first pupil from each group to roll the die and decide where to write the rolled number on his or her game card.
2. Turns are taken in a clockwise direction until every pupil has filled each of the boxes on his or her game card, one digit at a time.
3. Each pupil may "pass" on one of the rolls. One good reason to pass, for example, would be if a pupil rolled a 3 and a 2 on his or her first two rolls and placed them in the second and third boxes of the game card, respectively. If the third roll comes up as a 1, the pupil may opt to pass in the hope of

Activities for Elementary School Mathematics

      obtaining a higher number on the fourth roll.
4. Circulate among the pupils, checking for understanding, and giving assistance as necessary.

> **NOTE:** *Caution the players prior to beginning the game that numbers cannot be moved or changed once they have been written on the game card, nor may a player change his or her mind about "passing" on a number, once they have decided to do so.*

Closure or Evaluation:
    The winner is the pupil from each group with the largest three-digit number.

Page 2.18   Activities for Elementary School Mathematics

SUM WAR

Topic: PLACE VALUE

Grade Level: 2-4   **Activity Time:** 1 class period

Materials Needed:
1. A deck of fifty 3 x 5 index cards (five sets of cards with a numeral ranging from 0 to 9 on each of them) for each group of two to four pupils
2. Paper and pencil for each pupil

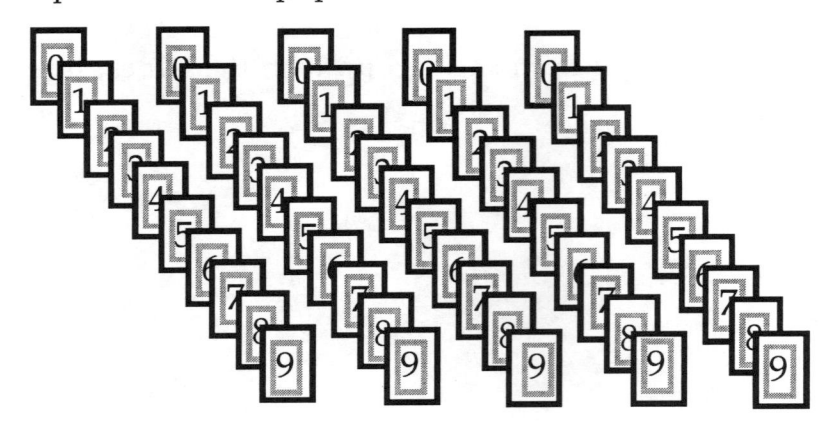

Objectives:
As a result of this activity, the learner will:
- recognize larger numbers.
- be able to find the sum of two-digit numbers.

Introduction:
1. Divide the pupils into groups of two to four.
2. Shuffle each deck of cards and place one deck in the middle of each group of pupils.
3. Tell the pupils that the object of the game is to create the largest number possible.
4. Each pupil draws one card to see who starts the activity (the highest cardholder starts, and the game proceeds to the right).

Major Instructional Sequence:
1. Instruct the first pupil from each group to take the top two cards from the deck and use them to create the largest number possible and record that number on his or her paper.

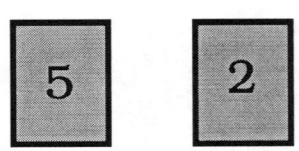

The pupil who draws a two and a five should choose the number 52 as the largest possible number.

Activities for Elementary School Mathematics

2. The next players in each group follow the same procedure, and the round continues until each player per group has created a two-digit number.
3. Circulate among the pupils, check for understanding, and give assistance when appropriate.
4. Play continues until each pupil has four or five two-digit numbers on his or her piece of paper.

Closure or Evaluation:
1. Instruct the pupils to add their lists of numbers.
2. The winners are the players with the highest sum from each group. (You may check addition with a calculator to confirm the sum of each group's winner.)

PLACE-VALUE "GO FISH"

Topic: PLACE VALUE / EXPANDED NOTATION

Grade Level: 4-6   **Activity Time:** 1 class period

Materials Needed:
1. Sets of 3 x 5 cards, three cards per set, each card in a set containing a different expanded notation of the same number. (Provide one deck of 60 cards, consisting of 20 sets, per each group of four pupils.)

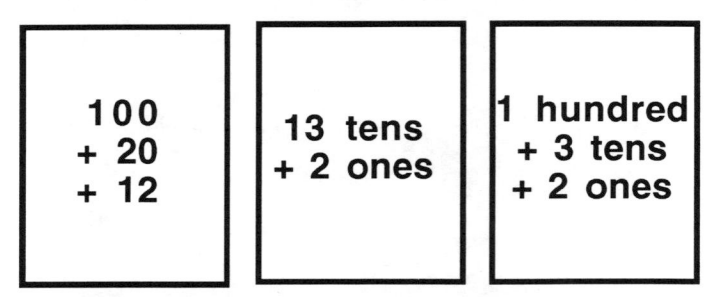

```
┌─────────┐  ┌─────────┐  ┌──────────┐
│  1 0 0  │  │ 13 tens │  │1 hundred │
│  + 20   │  │ + 2 ones│  │ + 3 tens │
│  + 12   │  │         │  │ + 2 ones │
└─────────┘  └─────────┘  └──────────┘
```

Each of the cards in this set represents an expanded notation of the number 132.

2. Paper and pencil (for scorekeeping), one each per group of four

Objectives:
As a result of this activity, the learner will:
- use expanded notation with words or numerals to understand place value.

Introduction:
1. Divide pupils into groups of four.
2. Tell the players from each group to select a card to see who is first dealer. The player from each group who has the highest card drawn becomes the dealer. From then on play moves to the right of the original dealer.
3. Instruct the dealers to shuffle the cards and pass them out to themselves and to the other three members of their groups. Each pupil should receive 15 cards.
4. Explain that the object of the game is to look for sets of three cards having the same numerical value.
5. Ask any player who finds that he or she has a set of three cards having the same value to lay the cards at his or her side, face up, on the playing surface. Tell the players that during the game they should lay down matching sets of three cards as soon as they complete a set. Then the game is ready to begin.

Activities for Elementary School Mathematics

Major Instructional Sequence:
1. Instruct the dealer to ask any of the other players in his or her group for a card, trying to complete a partial set that the dealer may have in his or her hand. If the queried player has a card of the value requested, he or she must give it to the dealer. (For example, if trying to complete the set of three cards in the previous illustration, the dealer would ask, "Do you have a card that equals 132?")
2. If the dealer is successful in obtaining a card, he or she makes a request for another card from any player in the group. This process continues for as long as the dealer keeps obtaining the requested card. If at any time a player does not have the card that the dealer requests, that player tells the dealer to "go fish," and the player to right of the person who requested the card takes a turn.
3. Circulate among the pupils, checking for understanding, and giving assistance as necessary.
4. Play continues until a player has no more cards in his or her hand.

Closure or Evaluation:
1. Players receive three points for each set of three matched cards they have placed face up on the table.
2. One point is deducted for each card that a player has left in his or her hand at the game's end.
3. The winner from each group is the player with the most points.

Activities for Elementary School Mathematics   Page 3.1

CHAPTER THREE
ADDITION • SUBTRACTION

ADDITION

Addition can be taught using several different approaches. One of these approaches is the *union of disjoint sets model,* putting two or more sets of separate items together to form a new set, called the sum. A set of 3 red blocks joined with a set of 2 blue blocks to form a set of 5 total blocks is addition. The addition sign (+) and the equal sign (=) can be included during the operation or added later.

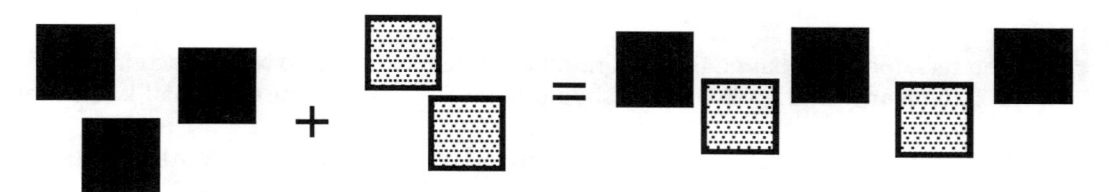

Pupils need to have hands-on experiences putting sets together. Rationally counting each set before they are joined together and then counting the total once the sets have been joined is essential to help pupils understand addition. Numeral cards and sign cards can be used to help understand the concept and connect it to the number sentence.

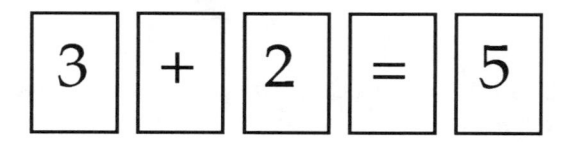

$$3 + 2 = 5$$

A second model of addition is the *linear approach.* A number line can be used to illustrate addition. The same number sentence would be illustrated as follows using the linear model:

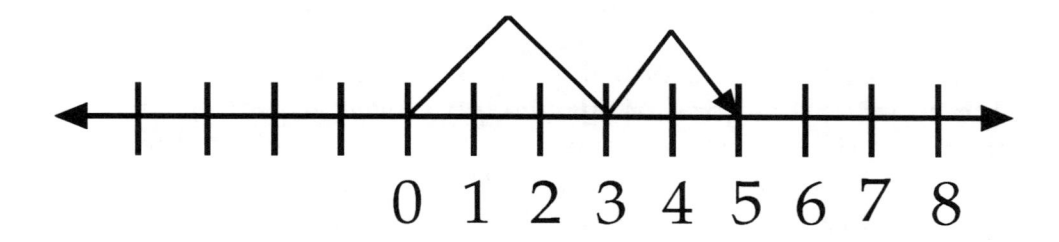

The number line could be attached to a pupil's desk and the pupil could move his/her finger along the line to illustrate the number sentence. A walk-on number line could also be used to illustrate the problem. (See page 1.22 for description.)

On the walk-on number line the pupil walks and rationally counts each step to 3, then takes 2 more steps and stops on 5. This is proof that 3 + 2 is equal to 5.

The different properties of addition need to be considered as an aid for helping pupils learn the basic addition facts. The *commutative property* is one of the most important in learning the basic facts. Pupils learn the concept long before they learn its name.

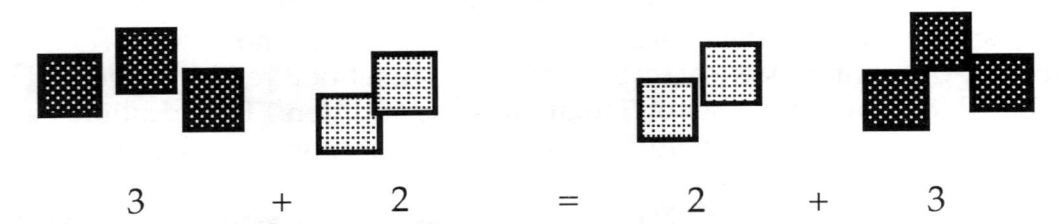

$$3 \quad + \quad 2 \quad = \quad 2 \quad + \quad 3$$

Rearranging the sets does not change the total number of items in the two sets, which in this example is 5. If the pupil learns that 3 + 2 = 5 they should know 2 + 3 = 5 because the total is the same.

The *identity property* states that when you add 0 to a number you get the same number. Fractions, decimals, etc. all conform to this property, N + 0 = N. Children learn quickly that 0 plus anything is the number being added to zero.

The *associative property* states that changing the grouping does not change the result. (3 + 2) + 1 = 3 + (2 + 1), each yielding a total of 6.

In learning the basic facts for addition the associative property is the least useful, but it will become very beneficial later on when 3 or more addends are used.

SUBTRACTION

The introduction of subtraction can be accomplished through a number of different techniques. Two ways that are more appropriate for early grades are considered here.

The most common method of introducing subtraction to young children would be *take-away* subtraction. Just as addition may be introduced as putting sets of objects together (the union of disjoint sets), subtraction may be taught as the taking apart of sets — the inverse operation of addition. To illustrate a take-away problem, use a set of 6 objects and remove (take away) 2 objects from the set of 6, which leaves 4 objects.

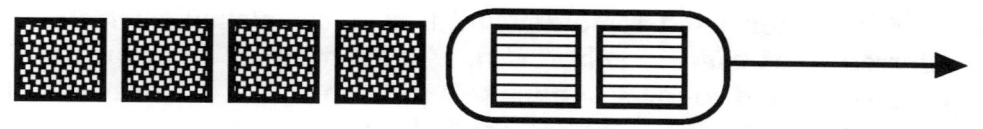

Activities for Elementary School Mathematics   Page 3.3

"6 minus 2 leaves 4" is the illustration in take-away form.

The other method which works well with the younger grades is *comparison subtraction*. The same problem, 6 – 2 = 4, would look as follows:

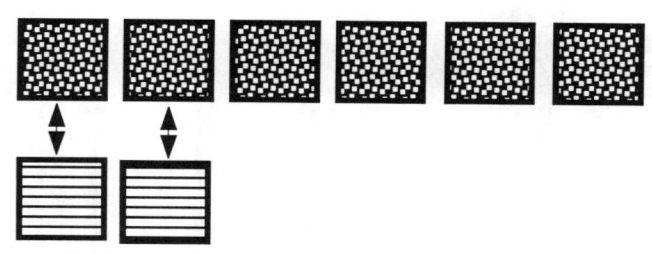

2 are matched (associated) but 4 are not matched (unassociated).

A comparison of the two sets is necessary to find the answer. The set of 2 is compared, or matched, to the set of 6. There are 4 left unassociated, or not paired, with any of the squares in the first set. Therefore, **6 - 2 = 4** is also the equation for the comparison problem illustrated above.

The following story problems may help clarify the difference in the two examples given above. **Wendi had 6 apples and gave 2 of her apples to Kristi. How many apples did Wendi have left?** This is an illustration of a take-away problem. After giving 2 of the 6 apples away, Wendi had 4 left.

Jane had 6 apples and Terry had 2 apples. How many more apples did Jane have than Terry? This is not a take-away situation, since there are two sets of apples and none of Jane's apples were removed or given to Terry. A comparison of the two sets is necessary to find the answer.

Other methods to illustrate subtraction are more appropriate for later elementary grades. One such method is *part-whole subtraction,* sometimes referred to as the *subset model of subtraction*. Consider the following example of the part-whole subtraction model: **There is a set of squares which contains 6 squares, two of them are striped and the rest are not striped. How many are not striped?** The total number in the set is known and a subset is also known. The whole is 6 and the known part, or subset, is 2. What is the unknown part of the set of squares? A drawing of this situation may help solve the problem.

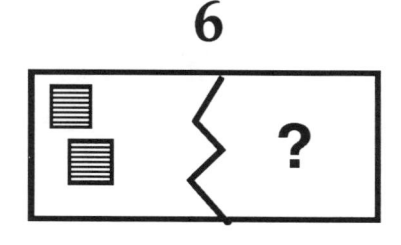

Another method often used for older children is *completion subtraction*, sometimes referred to as *additive subtraction*. An example of an appropriate story problem using this method follows: **To make a pie, 6 peaches are needed. If the cook has 2 peaches, how many more are needed?** The number of peaches needed to make a pie or to complete the set is 4. This would be illustrated differently than the other examples.

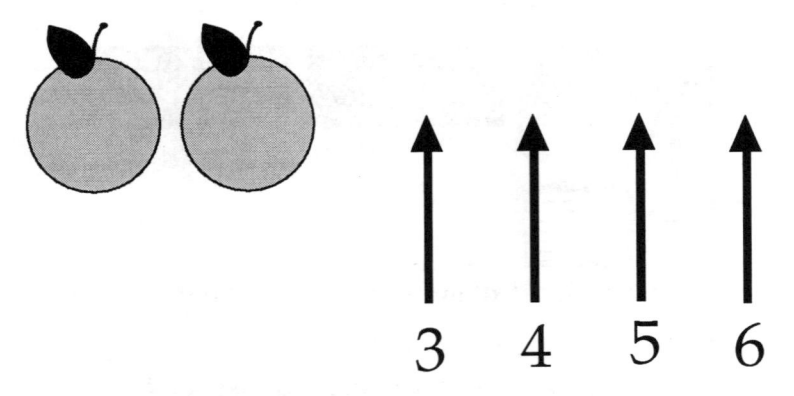

Four peaches are needed to complete the set.

Each of these four methods of illustrating subtraction requires a different approach when using manipulatives. The following activities provide children with experiences in all four methods of subtraction.

Activities for Elementary School Mathematics

+ & − CONCENTRATION

Topic: BASIC ADDITION AND SUBTRACTION FACTS

Grade Level: 1-3

Activity Time: 1 class period

Materials Needed:
1. Two sets of cards fifteen cards, one set with addition and subtraction facts, the other set with the answers, for each group of 3 to 4 pupils
2. Paper and pencil for each pupil in class
3. A flat playing surface, such as a large desk or table, for each group

3 + 2	5 − 2	6 + 6	10 − 3	6 + 3
5	3	12	7	9

There should be fifteen problem cards and fifteen answer cards for each group.

Objectives:
   As a result of this activity, the learner will practice and master:
   • basic addition and subtraction facts.

In preparation for this activity, set up a CONCENTRATION game board with five rows of six cards placed face down for each group of pupils.

Introduction:
1. Divide the class into small groups of three or four pupils.
2. Tell the class that they are going to play CONCENTRATION.

Major Instructional Sequence:
1. Explain to the pupils the process of searching for matches on the board: Each pupil in the group will take a turn looking for matches by turning over any two cards at a time. If the cards match (for example 7+5 and 12, or 5–2 and 3), the pupil who finds that match keeps the cards and play moves to the next player. If no match is made, the cards are returned to their original face-down position on the playing surface and the next pupil takes a turn. (Emphasize the importance of remembering the location of the cards in order to make a match.)
2. Begin the game.
3. Circulate among the groups of pupils, checking for understanding and giving assistance as necessary.
4. The game continues until all of the cards have been matched.

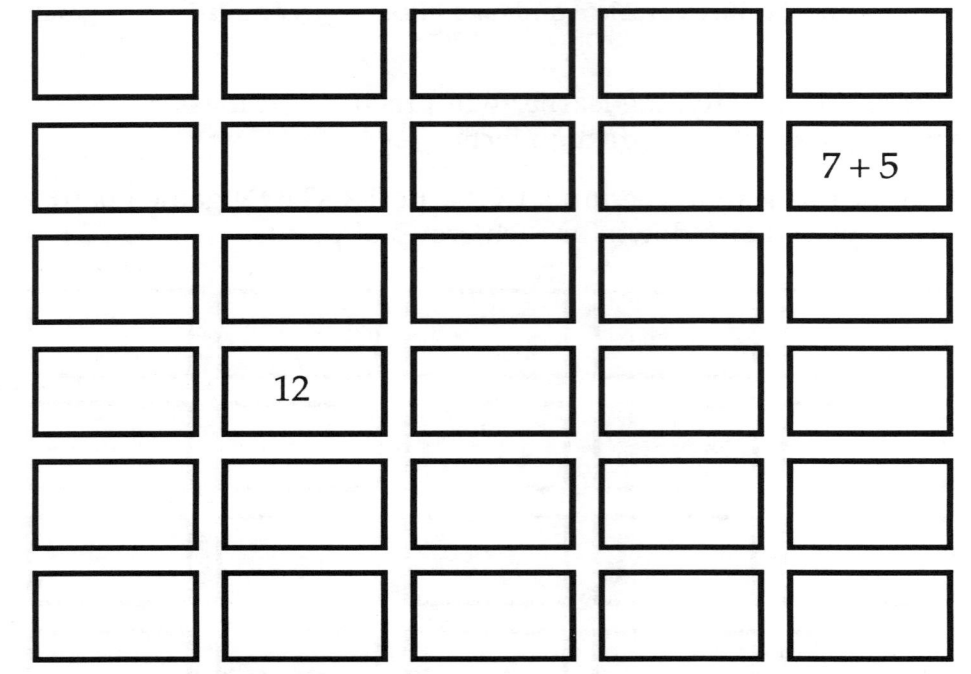

In this example, the pupil has made a match, 12 and 7+5.

Closure or Evaluation:
Play continues in each group until all cards are matched. The winner is the one who has the most cards.

Variation:
The game can be expanded to include multiplication and division.

Activities for Elementary School Mathematics

HOW MANY WAYS

Topic: BASIC ADDITION AND SUBTRACTION FACTS

Grade Level: 1-2 **Activity Time:** 1 class period

Materials Needed:
1. Nine white beans and nine dark beans for each group of pupils

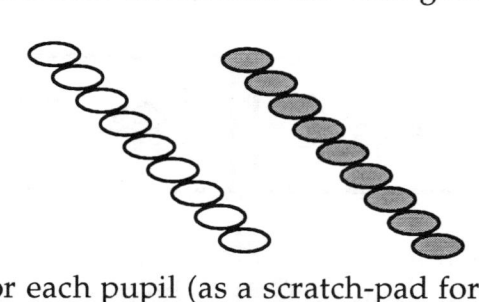

2. Paper and pencil for each pupil (as a scratch-pad for problem solving)
3. Overhead projector and screen (for groups to show their findings to the rest of the class)
4. Clear transparency and water-soluble marking pens for each group

Objectives:
As a result of this activity, the learner will find:
- the sum family for a given number.
- all the subtraction combinations for a number.
- related facts.

Introduction:
1. Divide the class into groups of three or four pupils.
2. Give each group the same number of beans (for example, 6 white and 6 dark).
3. Explain that the object of the game is to find as many subtraction combinations as possible for the number of beans they received by subtracting one color from the other, as in the examples below. (If addition is being taught, the object is to find as many sum families as possible.)

Major Instructional Sequence:
1. One group member illustrates a subtraction combination for the number of beans given to the group.

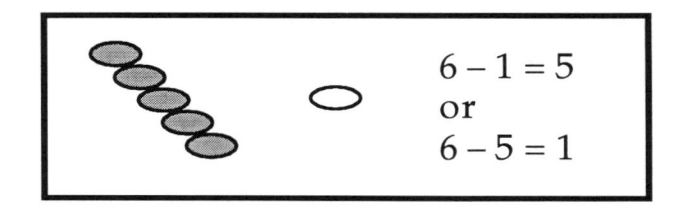

$6 - 1 = 5$
or
$6 - 5 = 1$

2. The next pupil in the group illustrates another combination.

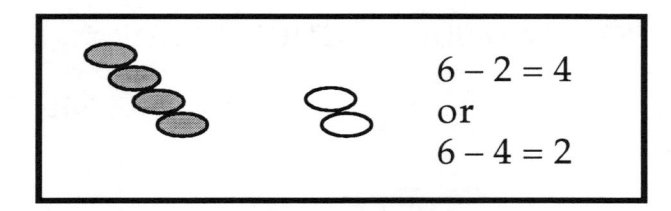

$$6 - 2 = 4$$
or
$$6 - 4 = 2$$

Other examples:

$$6 - 3 = 3 \qquad 6 - 0 = 6$$

3. The round continues until all of the combinations for six have been found. Pupils may use related addition facts to help them solve subtraction problems.
4. By drawing and coloring their combinations on a transparency, pupils use the overhead to show and discuss their combinations with the rest of the class.
5. Circulate among the groups, checking for understanding, and giving assistance when appropriate.

Closure or Evaluation:
Select different amounts of beans to explore other subtraction combinations and/or sum families.

Activities for Elementary School Mathematics

SUB TOSS

Topic: BASIC SUBTRACTION FACTS

Grade Level: 1-2 **Activity Time:** 1 class period

Materials Needed:
1. For each group of 4 pupils: a shoe-box lid, 2/3 of the inside painted blue and 1/3 painted red, with a small white circle painted for a target area (as shown), and 18 beans for tossing (or dropping) into the lid

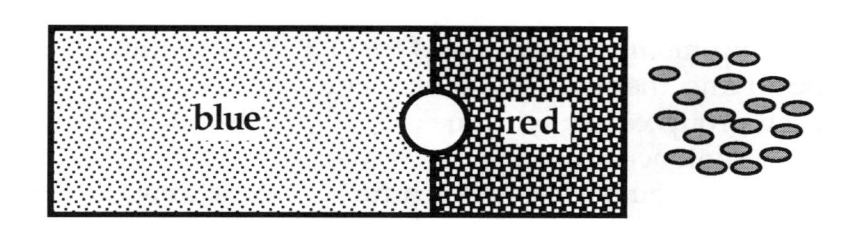

2. Paper and pencil for each four-member group for writing a subtraction equation representing beans tossed into the lid

Objectives:
As a result of this activity, the learner will:
- learn basic subtraction facts.
- write an open number sentence for a given subtraction problem resulting from a toss.

Introduction:
1. Separate pupils into groups of four.
2. Select a number of beans for the first round, for example, six.
3. Each pupil is given six beans for the first round of the activity.
4. Select a number from 1 to 20 and have pupils guess the number. The pupil in each group who guesses closest to the number goes first and play proceeds to the right.
5. Pupils take turns tossing or dropping the selected number of beans into the shoe-box lid, aiming at the white target area, and being careful that all of the beans stay inside of the lid. (Beans landing inside the target area are judged by the pupil doing the tossing to be on either the blue side or the red side because the target area straddles the line.)

Major Instructional Sequence:
1. Each pupil writes a number sentence (basic subtraction fact) representing

the result of his/her toss.
2. The number sentence is based on the total number of beans minus the number landing in the blue area which equals the number landing in the red area. An example for **6 - 4 = 2** follows:

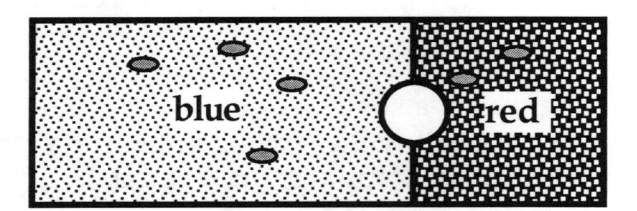

3. The other group members take turns tossing the same number of beans as the first pupil and then write number sentences.
4. Monitor each group carefully to make sure pupils are writing and answering their number sentences correctly.
5. After each group member has taken a turn, select a new number of beans for the second round.

Closure or Evaluation:
1. Activity continues as time permits.
2. Winners are all the pupils that write and answer their subtraction number sentences correctly.

Variation:
Pupils could also write a subtraction number sentence for the number of beans that fall into the red areas. **6 - 2 = 4** would be the correct sentence for the example above.

Activities for Elementary School Mathematics

WALK-ON NUMBER LINE ADD / SUB

Topic: BASIC ADDITION FACTS

Grade Level: 1-2

Activity Time: 1 class period

Materials Needed:
1. A walk-on number line

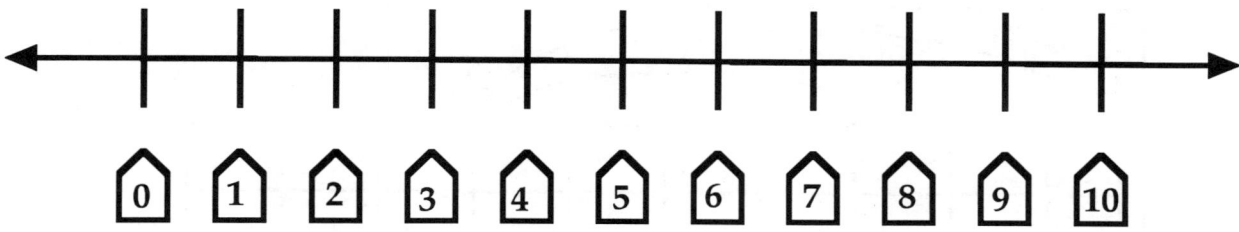

The walk-on number line can be made of heavy plastic, such as an old shower curtain. Affix the number line to the floor with masking tape, using black tape for the center line and cross lines, spaced so pupils can walk on it. The numbers should be placed on the floor next to the number line, with their arrow-like sides pointing to each appropriate intersecting line.

2. Cards (as illustrated above) with numerals ranging from 0 through 10, made of heavy card stock and laminated for endurance (For larger groups of children, make two sets of cards, one set for each side of the number line. Children can then walk on both sides of the center line.)

Objectives:
As a result of this activity, the learner will be able to:
- add basic facts using numbers less than ten.
- subtract with numbers less than ten.

Introduction:
1. To begin the activity, allow pupils to simply walk down the number line, with no counting involved.
2. Next have pupils walk and rationally count as they do so.

Major Instructional Sequence:
1. Start with addition facts having sums less than ten, for example, 3 + 2.
2. Instruct the first pupil to walk down the number line to the 3, then take two more steps (ending up at the 5).
3. Have the pupil count each step out loud, for example, "one, two, three," and then, "one, two," then finally calling out, "Three plus two equals five." The answer is verified by the number card where the pupil stopped.
4. The next pupil works a different problem, such as 4 + 6, and play continues

Page 3.12   Activities for Elementary School Mathematics

in this fashion until all pupils have had several opportunities to walk out a problem on the number line.

Note: *There are two ways the number line can be used for subtraction. Look at the following examples for using the number line to solve 7 − 4.*

The pupil starts at zero and walks to the 7, then turns around and walks back four steps, ending at the 3.

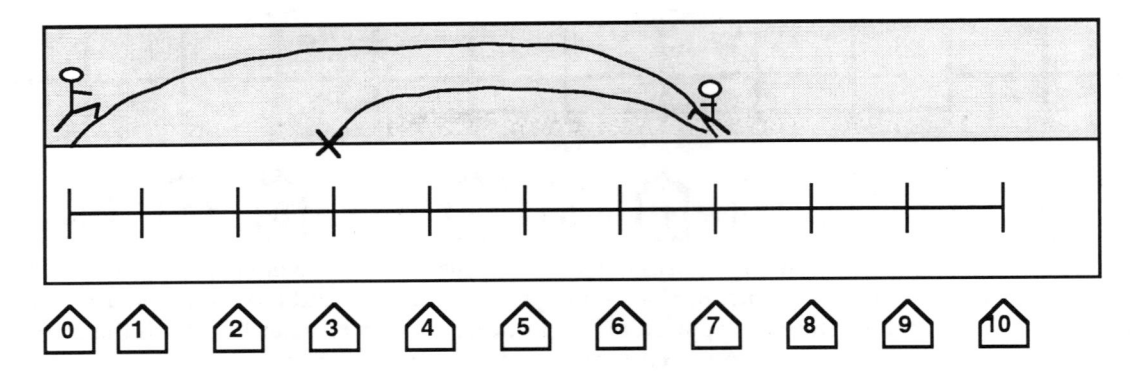

The pupil starts on the 7 and walks four steps back to the 3.

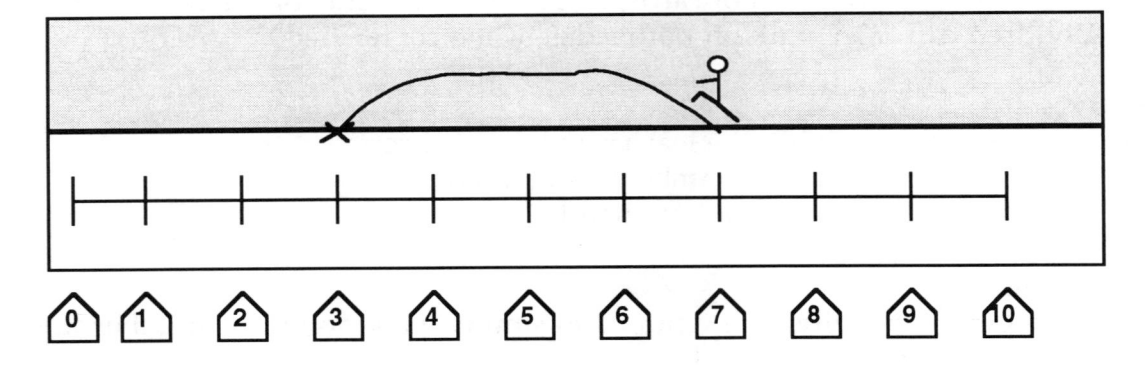

Closure or Evaluation:
Continue until each pupil has had an opportunity to walk out several problems in either addition or subtraction (or both, if appropriate) on the walk-on number line.

Activities for Elementary School Mathematics   Page 3.13

49 SUBTRACT

Topic: BASIC SUBTRACTION FACTS

Grade Level: 1-2　　　　　　　　　　　　　　**Activity Time:** 1 class period

Materials Needed:
1. A pair of number cubes, with numerals ranging from 1 to 18 on each of them, per group of four pupils (The range of numerals is dependent on the pupils' abilities and can be lower for younger children.)

2. A game board with 49 squares (7 X 7) for each pupil
3. Beans, buttons, etc. to be used as markers on the game board (49 markers per pupil).

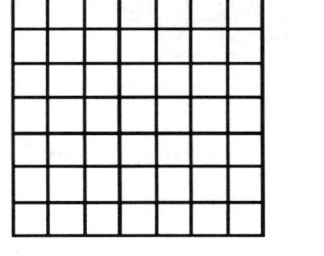

Objectives:
As a result of this activity, the learner will:
- recognize the larger of two numerals.
- subtract using basic facts.
- rationally count and remove markers for each difference in the numerals rolled on the die.

Introduction:
1. Divide the pupils into groups of four and distribute the game boards and markers.
2. Instruct the pupils to cover all 49 squares of their game boards with one bean per square.
3. Pupils in each group take a turn rolling one die to determine by highest score who goes first in the game. The pupil with highest roll on the die starts the game, and turns progress to the right.

Page 3.14   Activities for Elementary School Mathematics

Major Instructional Sequence:
1. The first pupil in each group rolls the dice.
2. Instruct pupils to subtract the smaller numeral from the larger. On the example below, the pupil would subtract 12 - 8 and get 4. (Pupils should be encouraged to do their subtraction work mentally. If they need help subtracting, counters could be used to assist them.)

3. Next, the pupil removes the number of markers from his or her game board represented by the difference in the two die, in this case 4.

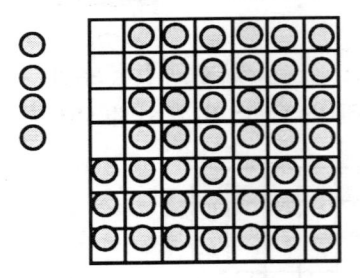

4. Circulate among the pupils, checking for understanding, and giving assistance when necessary.

Closure or Evaluation:
The players take turns until one of the players removes all of the markers from his/her game board. This player is declared the winner.

> **NOTE:** *As players have fewer markers remaining on their game boards, getting the necessary roll is more difficult. For example, if a player has only one marker remaining, he/she must roll, in turn, until a combination is rolled yielding a difference of one. The player rolls and passes until the right combination is rolled.*

Activities for Elementary School Mathematics   Page 3.15

ADDITION & SUBTRACTION TIC–TAC–TOE

Topic: BASIC ADDITION AND SUBTRACTION FACTS

Grade Level: 1-3   **Activity Time:** 1 class period

Materials Needed:
1. A three-by-three TIC-TAC-TOE card for each pupil (Cards could be laminated for repeated use by the pupils.)

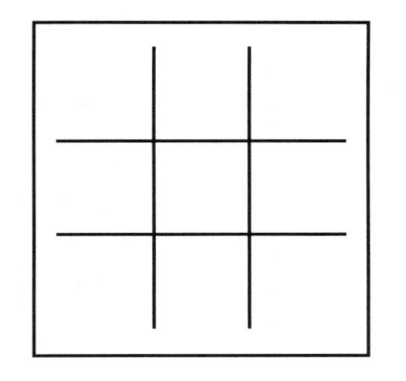

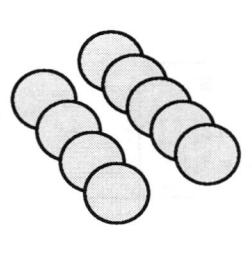

2. Chips to cover correct answers on game cards
3. Flash cards for addition and subtraction

Objectives:
As a result of this activity, the learner will:
- improve knowledge of addition and subtraction facts.

Introduction:
1. Make sure each pupil has a game card and 6 chips.
2. Have pupils write numerals from 0 to 18 at random in the squares of their game cards.

2	6	12
8	14	10
18	9	18

Page 3.16   Activities for Elementary School Mathematics

Major Instructional Sequence:
1. Call out one of the basic facts from an addition flash card.
2. Pupils check game cards to see if the answer is there and, if so, cover it with a chip.
3. Next, call out a basic fact from the subtraction facts cards; then alternate addition and subtraction facts until a winner emerges.
4. If you call out 4 + 2, for example, pupils will cover 6.

2	6	12
8	14	10
18	9	18

2	●	12
8	14	10
18	9	18

Closure or Evaluation:
1. The game goes on until a pupil has three in a row or column and is declared the winner. (Three answers in a diagonal may also be counted as a winner.)
2. Play new games as long as there is interest and a need. (Pupils may renumber their game cards for each new game if appropriate, or they may trade game cards.)

Activities for Elementary School Mathematics   Page 3.17

DICE ADDITION & SUBTRACTION

Topic: BASIC ADDITION FACTS AND SUBTRACTION

Grade Level: 1-3   **Activity Time:** 1 class period

Materials Needed:
1. A pair of dice for each group of three or four pupils
2. A pencil and paper for each pupil

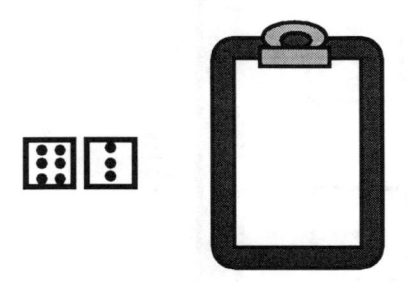

Objectives:
As a result of this activity, the learner will practice:
- basic addition facts.
- basic subtraction facts.

Introduction:
1. Pass out the dice to each group and have each pupil roll one die. The player with highest score in each group starts the activity.
2. Each player needs paper and pencil for score keeping. Instruct the pupils to set up their score sheets as follows:

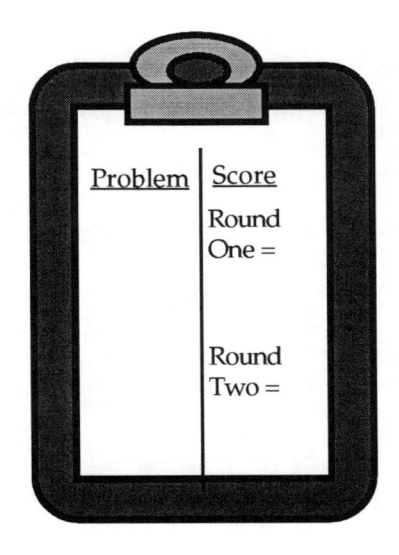

Problem	Score
	Round One =
	Round Two =

Major Instructional Sequence:
1. The first pupil in each group rolls the dice and records the results on his or her score sheet, placing the larger of the two numbers on top.
2. Instruct the pupils to subtract the smaller number from the larger number and record the resulting number in the score column.
3. Play continues until all have played and recorded five rounds.

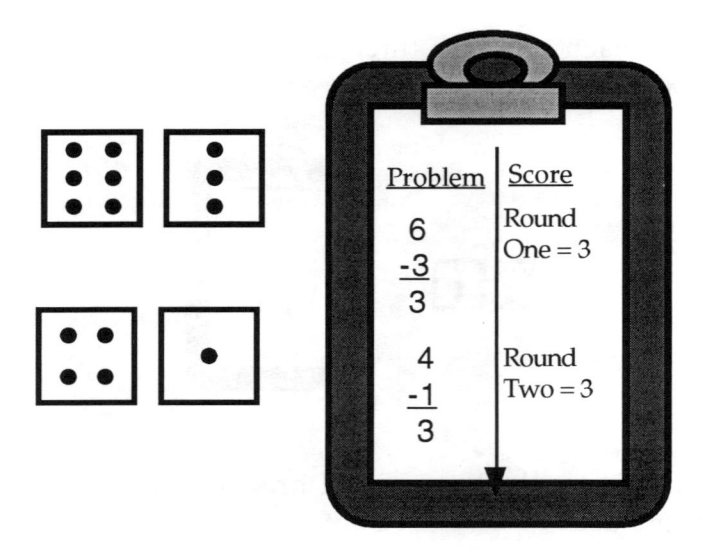

Closure or Evaluation:
1. Instruct pupils to add the scores of the five rounds to get a total score, and write their names on their score sheets.
2. In each group, the winner is the player with the highest total score.
3. Collect the score sheets to check for understanding.

Variation:
Use three or four dice to have larger subtraction problems. In such cases, subtract the lowest die from the higher two (or three).

Activities for Elementary School Mathematics

MARATHON ADD / SUBTRACT

Topic: BASIC ADDITION / SUBTRACTION FACTS

Grade Level: 1-3   **Activity Time:** 1 class period

Materials Needed:
1. Footprints with an addition or subtraction fact on the front of each footprint and the answer on the back

    9 - 5   4 + 5
      4   9

2. A notebook for each pupil

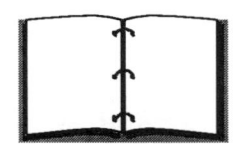

Note: *Pupils will need a notebook to keep track of the addition or subtraction facts they miss on their marathon facts run and practice any of the facts they miss. You may want to record the facts missed, too, so pupils can be helped later with unlearned facts.*

Objectives:
As a result of this activity, the learner will:
- become more proficient with basic addition and subtraction facts.

Introduction:
1. Select twenty-six of the footprints. (You may change the footprints as appropriate for pupils with differing levels of ability. Twenty-six footprints are used because marathons are 26.2 miles in length.)
2. Lay the footprints on the floor, problem-side up, in the order desired.
3. Explain the procedure as follows:
    - As you go down the footprints, look at the fact and say the fact and answer aloud.
    - Pick up the footprint and verify your answer by looking on the reverse side.
    - If correct, replace the footprint and proceed to the next footprint and follow the same procedure.
    - If incorrect, record the fact and correct answer in your notebook for later study, then replace the footprint and proceed to the next footprint and follow the same procedure.

Major Instructional Sequence:
1. The first pupil goes down the twenty-six step run.
2. Monitor carefully and keep track of the steps missed, making sure the pupil records them in his or her notebook.
3. Pupils continue, in turn, to run the marathon and record missed facts in their notebooks.
4. All pupils go the distance.

Closure or Evaluation:
Pupils use their notebooks to review and practice missed facts.

Activities for Elementary School Mathematics

ROLLO-SUM/ROLLO-DIFFERENCE

Topic: BASIC ADDITION / SUBTRACTION FACTS

Grade Level: 1-3   **Activity Time:** 1 class period

Materials Needed:
1. A pair of dice for each pair of pupils (with numerals ranging from 1 to 9 on each die)
2. Paper and pencil for each pupil
3. A basic facts chart for each pair of pupils playing the game

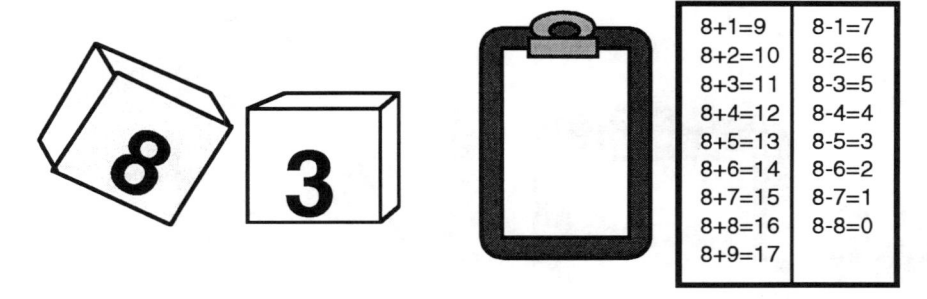

8+1=9	8-1=7
8+2=10	8-2=6
8+3=11	8-3=5
8+4=12	8-4=4
8+5=13	8-5=3
8+6=14	8-6=2
8+7=15	8-7=1
8+8=16	8-8=0
8+9=17	

Objectives:
As a result of this activity, the learner will master:
- basic addition and subtraction facts.

Introduction:
1. Pair each pupil with a partner.
2. Tell the pupils that they will work together in pairs to check each other on basic addition or subtraction facts as rolled by the dice.

Major Instructional Sequence:
1. Each pupil in a pair rolls a die to determine who is to start the game. (The highest roll starts.)
2. The first pupil in each pair rolls the dice and adds (mentally) the two numerals rolled on the dice. (For subtraction, the smaller of the two numerals is subtracted from the larger.)
3. The pupils' partners use the charts to check the answers.
4. If the sum or difference is correct, a point is received; if incorrect, the combination missed is recorded on paper for later practice. (Partners collaborate on whether or not a point is received, and each pupil records his or her own points and missed combinations on paper.)
5. The other partner in each pair rolls the dice and the same procedure is followed.

Closure or Evaluation:
The first pupil in each pair to get ten correct answers wins the round.

SUM / DIFFERENCE PENNY

Topic: BASIC ADDITION FACTS

Grade Level: 1-3

Activity Time: 1 class period

Materials Needed:
1. Twenty pennies per pupil
2. Place mats for each pupil (a sheet of construction paper or piece of cloth)
3. Paper and pencil for each pupil

Objectives:
As a result of this activity, the learner will:
- display knowledge of basic addition and subtraction facts.
- become familiar with sum families.
- become familiar with different combinations in subtraction for given numbers.

Introduction:
1. Instruct the pupils to sit on the floor in a large circle.
2. Put place mats and pennies in front of each pupil.

Major Instructional Sequence:
1. Each pupil flips the pennies onto the mat. (The pennies may be either "flipped" with the thumb and index finger or simply tossed onto the mat.)
2. Pupils count the "heads" resulting from their individual flips and write the numeral followed by a "+" sign.

3. Pupils count the "tails" and record the number after the "+" sign.
4. Pupils write the sum of the two addends after an "=" sign. (The result is a number sentence, such as 9+11=20.)
5. For subtraction, pupils count the number of "heads" and subtract from the total number of pennies (20 in this example), then write the appropriate number sentence, such as 20–9=11.
6. Circulate among the pupils, check for understanding, and give assistance as needed.
7. Use varying numbers of pennies for subsequent flips onto the mat.

Closure or Evaluation:
Have pupils flip the pennies at least five different times and create both addition and subtraction number sentences based on the outcomes.

OVERTURN/SPILL AND ADD OR SUBTRACT

Topic: ADDITION / SUBTRACTION

Grade Level: 1-3   **Activity Time:** 1 class period

Materials Needed:
1. Ten one-inch squares with a single digit, zero to nine, on each side (a different numeral on the back of each card), one set of numeral cards for each pair of pupils

front	back	front	back
7	4	9	6

   NOTE: *Sample graphics which may be reproduced on a copy machine are included in the appendix.*

2. An eight-ounce cup for each pair of pupils
3. Paper and pencil for each pair of pupils

Objectives:
As a result of this activity, the learner will:
- add one-digit numbers mentally.
- look for combinations that equal ten (which will make adding easier).
- become familiar with basic subtraction facts.

Introduction:
1. Divide pupils into pairs and distribute number squares and cups to each pair.
2. One pupil is to place number squares into the cup and the other pupil is to have paper and pencil ready to record the sum when the number squares are overturned.

Major Instructional Sequence:
1. The first pupil places his or her hand over the cup mouth and shakes the cup, then overturns the cup and spills the number squares onto the desk.
2. The second pupil mentally adds squares together and writes the sum of all the numbers on his or her paper.
3. The first pupil then gets paper and pencil, records the numerals from the desktop, and checks the addition.
4. If the second pupil did the mental addition correctly, he/she gets that many points for a score. If the sum is not correct, no points are awarded. (The

Activities for Elementary School Mathematics

      two pupils collaborate on the paper-and-pencil calculations.)
5. Circulate among the pairs, checking for understanding, and providing assistance as appropriate.
6. Players reverse roles and repeat the process.
7. For subtraction, only two number squares are used. After spilling the squares, the smaller numeral is subtracted from the larger numeral (mentally), and the partner records the number sentence for the subtraction. For the squares below, for example, the number sentence would be 9 – 7 = 2.

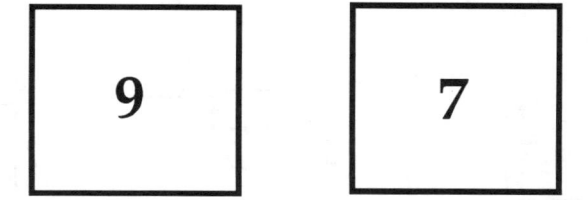

                              | 9 | | 7 |

Closure or Evaluation:
1. The game ends for addition when the sum of one pupil's scores in each pair reaches a predetermined number (like ninety), or the game can be timed.
2. For subtraction, the game ends when each pupil has successfully completed ten number sentences.

SUM/DIFFERENCE HOLD-UP

Topic: BASIC ADDITION AND SUBTRACTION FACTS

Grade Level: 1-3 **Activity Time:** 1 class period

Materials Needed:
1. Ten square pieces of card stock, each with one digit, zero to nine (one set of ten cards per pupil)

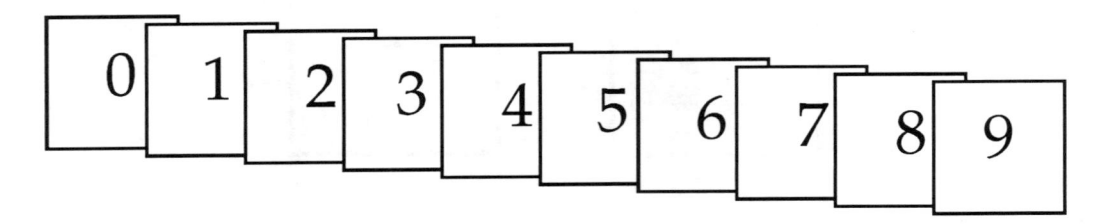

2. A prepared list of questions, such as:
    "What number is three more than six?"
    "If two is taken away from seven, how much is left?"
    "I am five more than three, who am I?"
    "Seventeen minus nine is how much?"
    "What is the answer when three, four and one are added together?"
    "Who am I if I am twice the answer you get when you add three and one?"

Objectives:
As a result of this activity, the learner will:
- display an understanding of the basic facts involved in addition and subtraction.

Introduction:
1. Pass out a set of ten cards to each pupil.
2. Have pupils arrange cards on their desks, face up, in order from zero though nine.

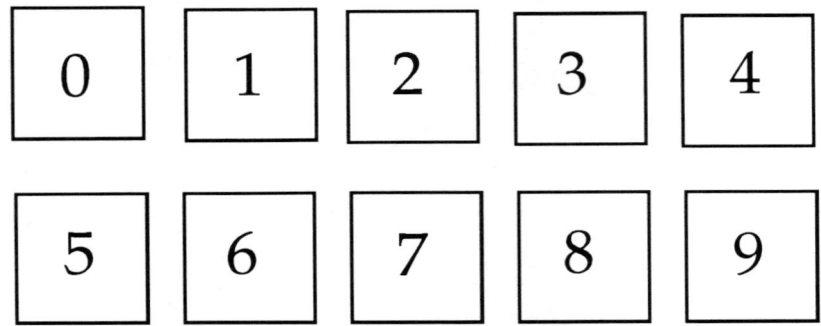

Major Instructional Sequence:
1. Read (or select a pupil to read) the first question, such as, "What do I get when I add three and two?"
2. Pupils will select their answers from the squares in front of them and hold up their cards.
3. Check for understanding by looking at all of the pupil responses. (Make a mental note of those pupils who may need additional instruction later.)

Closure or Evaluation:
After a set time period, the activity is stopped.

Variation:
In a separate activity, pupils could create questions for each of the ten cards. This would be an excellent collaborative activity for small groups.

Page 3.28   Activities for Elementary School Mathematics

DOMINO ADDITION / SUBTRACTION

Topic: BASIC ADDITION OR SUBTRACTION FACTS

Grade Level: 1-3   **Activity Time:** 1 class period

Materials Needed:
   1. One set of dominoes for the class

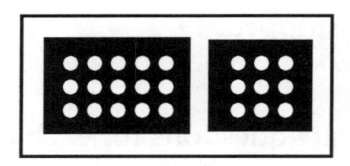

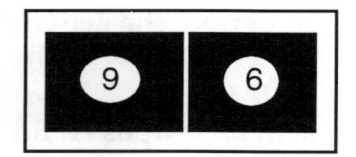

   The teacher could make a special domino set to cover higher subtraction facts if desired.

   2. Paper and pencil for each pupil

Objectives:
   As a result of this activity, the learner will become more familiar with:
   - number/numeral concepts.
   - addition facts.
   - subtraction facts.

Introduction:
   1. Put desks in a circle (or establish some route for the dominoes to travel from one person to another in a regular sequence).
   2. Furnish each pupil with one domino.

Major Instructional Sequence:
   1. Tell the pupils to turn the dominoes so that the larger number of dots (or numeral) is on top. (This will be helpful when doing subtraction.)

   $6 - 4 = 2$

   $6 + 4 = 10$

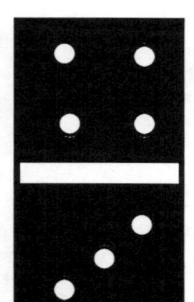

   $4 - 3 = 1$

   $4 + 3 = 7$

   2. Explain to the pupils that they are to write a subtraction and addition problem (number sentence) for each domino they receive.
   3. After they have written their problems, give a signal for pupils to pass their

Activities for Elementary School Mathematics

    dominoes along in the predetermined direction.
4. Upon receiving a new domino, each pupil creates new subtraction and addition problems (number sentences). A predetermined number of exchanges can be established for the game, or the game can continue until all of the dominoes have been passed around the class and returned to the point where they started.

Closure or Evaluation:
1. Have pupils volunteer to read their number sentences aloud, and the rest of the class gives a "thumbs-up" for correct number sentences and a "thumbs-down" for incorrect number sentences.
2. Ask pupils to circle any incorrect number sentences on their papers during this process.
3. Collect the pupils' papers so that you can check them for individual understanding.

Note: *If pupils find and mark the first number sentence (when it is read aloud) on their papers, the rest of the number sentences should appear in order, making the checking process easier.*

Page 3.30   Activities for Elementary School Mathematics

SHAKEDOWN

Topic: BASIC ADDITION FACTS FAMILY

Grade Level: 1-4   **Activity Time:** 1 class period

Materials Needed:
1. Plastic cup for each pair of pupils
2. Beans; some dark, some white (at least 10 dark and 10 white per pair of pupils)
3. Paper and pencil to record the outcome of each shakedown

Objectives:
As a result of this activity, the learner will become familiar with:
- addition facts.
- the addition facts family for a given number.
- subtraction facts.
- subtraction facts as they relate to addition.

Introduction:
1. Ask pupils a basic addition fact, such as 1 + 4. (Elicit 5.)
2. Next ask what 3 plus 2 equals. (Elicit 5.)
3. See if pupils can give other combinations of two numbers that equal 5.
4. If working with subtraction, ask pupils what 5 - 1 equals. (Elicit 4.)
5. Next ask what 6 minus 2 equals. (Elicit 4.)

Major Instructional Sequence:
1. Separate pupils into pairs.
2. Pass out one cup and a handful of beans (10 white and 10 dark) to each pair of pupils.
3. Have pupils place an assortment of white and dark beans into their cups (5 to 8 beans in the beginning; increase the amount for older pupils later).

4. Tell the pupils to *shake* the cups and dump the beans onto their desk tops.

5. Pupils collaboratively count and record the number of dark beans and the number of white beans on the desk top after the "shakedown."

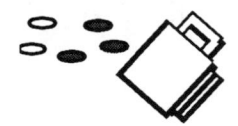

6. The pupils then collaborate to make a number sentence and find the sum. In the illustration below, there are three dark beans and two white beans; the number sentence and sum would be 3 + 2 = 5.

7. For subtraction, pupils count the total number of beans (five), then make a subtraction number sentence to take away the dark beans. In the illustration below, there are five total beans; the number sentence would be 5 − 3 = 2.

Closure or Evaluation:
Let pupils choose a new assortment of beans, put them into their cups, and follow the same procedures for a new "shakedown."

100 ADDER

Topic: ADDITION OF TWO-DIGIT AND ONE-DIGIT NUMBERS

Grade Level: 1-6   Activity Time: 1 class period

Materials Needed:
1. A large hundreds chart for the front of the room (see illustration on the following page)
2. Individual hundreds charts for each pupil or for each pair of pupils

Objectives:
As a result of this activity, the learner will:
- be able to add one- or two-digit numbers.
- become familiar with number patterns in addition.

Introduction:
1. Display the hundreds chart and discuss patterns that the children see on the chart.
2. Pass out the pupils' charts, permit pupils time to explore any patterns previously mentioned, and discuss any new patterns they see.

Major Instructional Sequence:
1. Call out a basic fact combination, such as 5 + 7, and have pupils place a finger on the first number called out, then count to the second number in the combination. (Pupils would place a finger on 5 and count seven spaces to the 12.)
2. Circulate around the room and check for accuracy.
3. Next, ask pupils to work 25 + 7. (Help pupils notice the pattern: the answer, 32, ends in a 2 just like the 5 + 7 combination before.)
4. 85 + 7 could be next, with the idea that the answer also ends with a 2.

Closure or Evaluation:
Explore other similar patterns on the hundreds chart.

Activities for Elementary School Mathematics

HUNDREDS CHART

1	2	3	4	5	6	7	8	9	10
11	12	13	14	15	16	17	18	19	20
21	22	23	24	25	26	27	28	29	30
31	32	33	34	35	36	37	38	39	40
41	42	43	44	45	46	47	48	49	50
51	52	53	54	55	56	57	58	59	60
61	62	63	64	65	66	67	68	69	70
71	72	73	74	75	76	77	78	79	80
81	82	83	84	85	86	87	88	89	90
91	92	93	94	95	96	97	98	99	100

Page 3.34   Activities for Elementary School Mathematics

SUM / DIFFERENCE WAR

Topic: PLACE VALUE, ADDITION AND SUBTRACTION

Grade Level: 1-3   **Activity Time:** 1 class period

Materials Needed:
1. Sets of cards with the numerals zero to nine on them (one set per each group of four pupils)

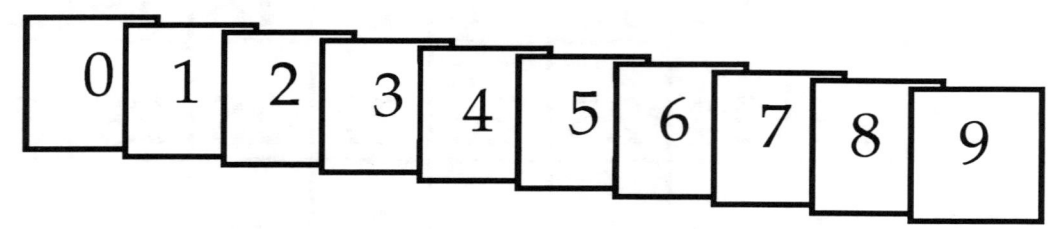

[Illustration of cards numbered 0 through 9]

2. Paper and pencil for each pupil

Objectives:
As a result of this activity, the learner will:
- recognize larger numbers.
- be able to find the sum of two-digit numbers.
- be able to find the difference of two numbers.

Introduction:
1. Divide pupils into groups of four.
2. Shuffle the cards and place a stack with each group.
3. In each group, pupils draw a card to see who starts the activity. The highest numbered card starts, and the turns progress to the right.

Major Instructional Sequence:
1. Instruct the first pupil from each group to take the top two cards from the stack and make the largest two-digit number he or she can think of for the cards drawn. For example, eighty-two would be the best choice below.

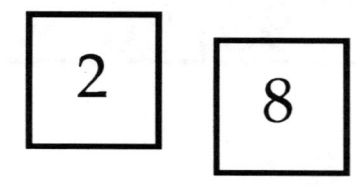

[Illustration of cards showing 2 and 8]

2. The pupil records the number on his/her paper.
3. The next player does the same, and the round continues until each player has a two-digit number.

Activities for Elementary School Mathematics

    4. Play continues until each pupil has four or five numbers recorded.

Closure or Evaluation:
1. Pupils add their lists of numbers.
2. The winner is the person with the highest sum.

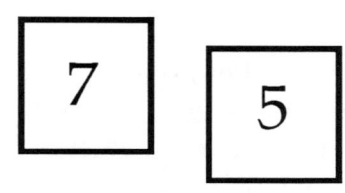

$$\boxed{7} \quad \boxed{5}$$

Two would be the correct response for these cards, 7 - 5 = 2.

Variation:
1. For subtraction, pupils turn up two cards and subtract the smaller number from the larger one. The difference is recorded.
2. The game continues for a set number of rounds. Pupils add their answers, and the winner is the pupil with the lowest total. Calculators could be used to verify totals.

Page 3.36 Activities for Elementary School Mathematics

SUBTRACTION IS LAST

Topic: BASIC ADDITION FACTS AND SUBTRACTION

Grade Level: 1-3 **Activity Time:** 1 class period

Materials Needed:
1. Three number cubes with numbers selected by the teacher to meet the needs of the pupils (Each group of players should have three cubes.)
2. Paper and pencil to record the score of each player (one scorekeeper per group)

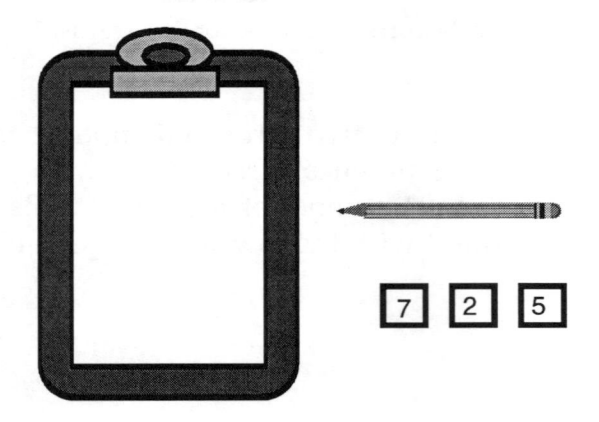

Objectives:
As a result of this activity, the learner will:
- learn basic addition facts.
- use subtraction facts.

Introduction:
1. Divide pupils into groups of four.
2. Each group selects a scorekeeper who writes each player's name on the score sheet.
3. Pupils roll a number cube to determine who is to start the game in each group. The pupil with the highest number goes first.

Major Instructional Sequence:
1. Instruct the first player in each group to roll the three number cubes, add the two larger numbers together, and then subtract the smallest number from the sum.

With this roll, the pupil would add the 7 and the 5, get a sum of 12, then subtract 2 from 12 for a score of 10.

2. Circulate among the pupils, checking for understanding, and providing assistance where needed.
3. Play continues to the right of the first player, and in the same manner.
4. The scorekeeper records each player's score, which is collaboratively agreed upon by the rest of the group. If a player adds or subtracts incorrectly, his/her score for that round is zero.

Closure or Evaluation:
1. Instruct pupils to add up their scores from the round.
2. The game continues for a specified number of rounds or until a preset time limit is reached.
3. The winner is the pupil with the highest total in his or her group at the end of the game.

ADDITION/SUBTRACTION BINGO

Topic: BASIC ADDITION OR SUBTRACTION FACTS

Grade Level: 2-4 **Activity Time:** 1-2 class periods

Materials Needed:
1. BINGO game cards, one per pupil
2. Beans to use as chips in the BINGO game

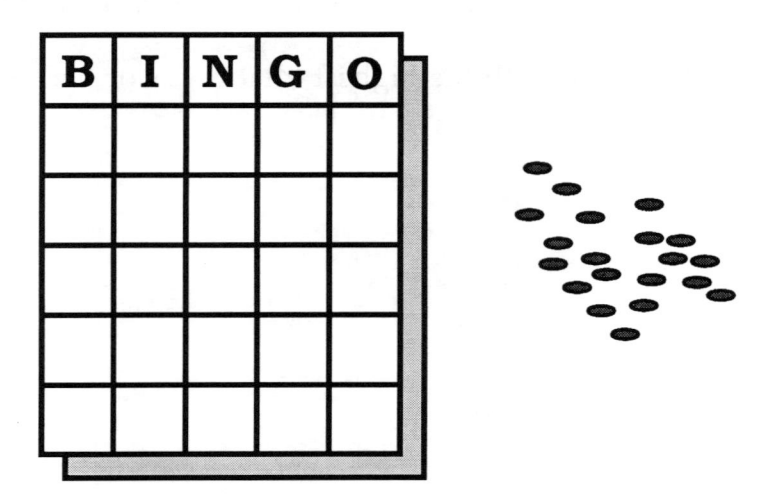

Objectives:
As a result of this activity, the learner will become more familiar with:
- basic addition facts.
- basic subtraction facts.

Introduction:
1. Pass out the BINGO cards and beans to pupils.
2. Instruct pupils to write numerals in the boxes on their cards which might represent answers to addition and subtraction facts to be called out from flash cards during the BINGO game.
Demonstrate with an example: "One of the flash cards says 3 + 5, so I know 8 would be a good number to place on the BINGO card. One of the flash cards says 7 − 4, so I know 3 would be a good number to place on the BINGO card."
3. Display flash cards on the chalktray so that pupils will have appropriate models from which to choose.

Major Instructional Sequence:
1. Mix addition and subtraction flash cards together, shuffle, and place in a stack.
2. Select the first flash card and by randomly selecting one of the letters in BINGO, say, "Under the B, 4 + 2." Pause for pupils to calculate combinations

Activities for Elementary School Mathematics   Page 3.39

mentally and place markers on their BINGO cards if the answer is there. Then select the next flash card, randomly select another letter in BINGO, and say, "Under the N, 7 – 5," and so on.

Note: *Write the selections down in a list as they are called, for later checking.*

B	I	N	G	O
13		10	19	3
2	5			6
1	16	8	17	11
18	14	5	3	7
	8	15	18	9

4. Five beans in any direction (down, across, or diagonally) is a BINGO.

Closure or Evaluation:
The game continues until a winner emerges by calling out "BINGO!" and his/her card is checked for accuracy from the calling list you have kept.

ADDITION/SUBTRACTION BINGO II

Topic: BASIC ADDITION / SUBTRACTION FACTS

Grade Level: 2-4   **Activity Time:** 1 class period

Materials Needed:
1. BINGO game cards, one per pupil, with addition and subtraction combinations
   Note: *The combinations may be placed on the cards in a separate activity where small groups of pupils use flash cards to drill on the combinations as they place them on the cards.*
2. Beans to use as markers or "chips" in the BINGO game

B	I	N	G	O
3+4	1+5	7-3	8+3	9-4
3+0	6-2	8+6	12-3	8-0
2+1	15-5	2+9	18-9	8+8
16-8	9+9	13-9	17-7	4+9
19-3	6+5	4-3	11-1	6+7

Objectives:
As a result of this activity, the learner will:
- become more familiar with basic addition and subtraction facts.

Introduction:
1. Pass out the BINGO cards and beans to pupils.

Major Instructional Sequence:
1. Mix addition and subtraction flash cards together, shuffle, and place in a stack.
2. Select the first flash card and by randomly selecting one of the letters in BINGO, say, for example, "Under the B, find a combination for 8." Pause for pupils to calculate combinations mentally and place markers on their BINGO cards if the answer is there. Then select the next flash card, randomly select another letter in BINGO, and say, for instance, "Under the N, find a combination for 3," and so on.

Activities for Elementary School Mathematics   Page 3.41

Note: *Write the selections down in a list as they are called so they can be checked later.*

B	I	N	G	O
3+4	1+5	7-3	8+3	9-4
3+0	6-2	8+6	12-3	8-0
2+1	15-5	2+9	18-9	8+8
16-8	9+9	13-9	17-7	4+9
19-3	6+5	4-3	11-1	6+7

3. Five beans in any direction (down, across, or diagonally) is a BINGO.

Closure or Evaluation:
The game continues until a winner emerges by calling out "BINGO!" and the winning card is checked for accuracy from the calling list you have kept.

Page 3.42   Activities for Elementary School Mathematics

RABBIT

Topic: BASIC ADDITION AND SUBTRACTION FACTS

Grade Level: 2-4   **Activity Time:** 1 class period

Materials Needed:
1. Sets of fifty-two game cards with addition or subtraction facts

   Note: *On half of the cards print addition and subtraction facts (combinations) and on the other half print the answers (one set per group of five pupils).*
2. On one card in each deck, draw a picture of a rabbit (or critter of your choice).

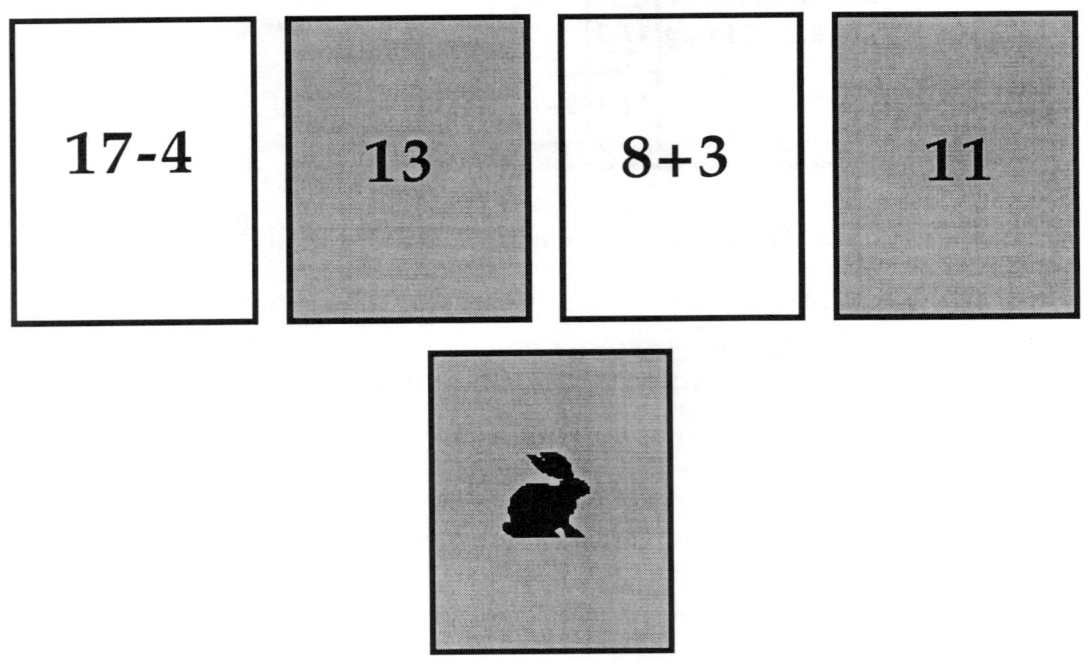

| 17-4 | 13 | 8+3 | 11 |

[rabbit card]

3. Paper and pencil for keeping score

Objectives:
As a result of this activity, the learner will:
- practice addition facts.
- practice subtraction facts.

Introduction:
1. Divide pupils into groups of five.
2. The card game is played like *Old Maid,* with the entire deck distributed to the five players.
3. Instruct the players to look through their hands for sums or differences that match the addition or subtraction facts they hold.

Activities for Elementary School Mathematics

4. The matched pairs are discarded face up on the table and checked by the other players for accuracy.
5. If an incorrect match is placed on the table, it is simply placed back in the player's hand without penalty.

Major Instructional Sequence:
1. When all of the matches from the original deal of the cards have been placed on the table, the player to the dealer's right selects a card from the player to his/her right.
2. If the player has a match for the card selected, the matches are discarded and the player continues to draw cards from the player to his/her right.
3. As long as the player keeps getting matches, the cards are discarded, and the player continues to select cards from the player on the right.
4. When a card has been selected and the player has no match, play moves to the right and continues in the same fashion.

Closure or Evaluation
1. Play continues until all matches have been made. The player holding the Rabbit card must subtract ten points from his/her score.

2. Scoring is calculated by adding the *answer* cards (answers to the matches laid down).

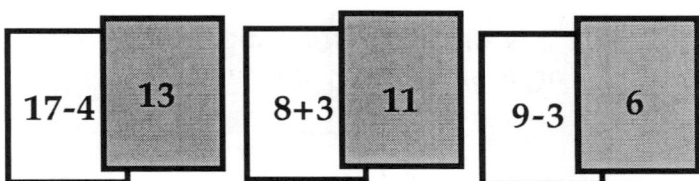

    17-4 13   8+3 11   9-3 6

    These cards would yield a score of 30: the sum of 13, 11, and 6.

3. The winner for each round is determined by highest score.

2 FOR 1 ADDITION/SUBTRACTION

Topic: BASIC ADDITION AND SUBTRACTION FACTS

Grade Level: 3-5 **Activity Time:** 1 class period

Materials Needed:
1. Decks of forty-eight cards, made from four identical sets of cards with the numbers from one to twelve on them (one deck per group of four pupils)

Objectives:
As a result of this activity, the learner will:
- match pairs of numbers to make selected sums.

Introduction:
1. Divide pupils into groups of four and select a dealer for each group.
2. The dealer passes out four cards to each player in the group, then places four cards face up in the middle of the group and places the deck to one side.

Major Instructional Sequence:
1. The player to right of the dealer begins the game. The object of the game is to find a card in his/her hand that is equal to two cards on the table.

> **Note:** *A match is made by either an addition combination or a subtraction combination.*

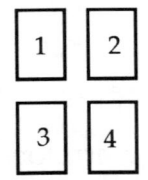

```
1  2
3  4
```

Matches may be made across or down. Some possible combinations for this illustration are: 1+2, 2+4, 3+4, 1+3, 2+1, 2-1, 4-3, 4-2, 3-1, for example.

Activities for Elementary School Mathematics

2. The player lays his/her card on the two-card combination which matches, and the group decides if the match is valid.
3. If the match is valid, the player collects the three cards and sets them aside and gets one new card from the deck.
4. Two new cards are placed in the middle of the table by the dealer to keep four cards face up at all times.
5. If a player cannot make a play, a card is discarded and a new one is dealt from the deck.
6. Play continues until the stack in the middle has been eliminated.

Closure or Evaluation:
The player with the most cards matched is the winner.

Activities for Elementary School Mathematics · Page 4.1

CHAPTER FOUR
MULTIPLICATION • DIVISION

MULTIPLICATION

Multiplication can be introduced in a number of ways, so that pupils can better understand what it actually means. A hands-on approach helps pupils understand the multiplication process. The *equivalent sets model* is one of the most commonly used methods of introducing multiplication. In this model a number of equivalent sets are used to represent multiplication. For example, 4 x 6 would look as follows:

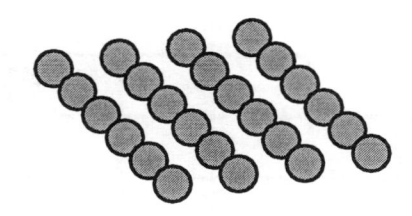

Four sets consisting of six items per set total twenty-four items. Similarly, 6 x 4 would be illustrated as follows:

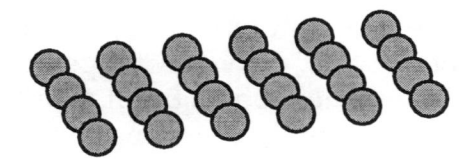

The commutative property could be used by pupils so that if they learned one of the examples illustrated above, they would not have to practice the other (6 x 4 = 4 x 6).

A *linear model* is often used to introduce multiplication. On the linear model, 5 x 2 would look like this:

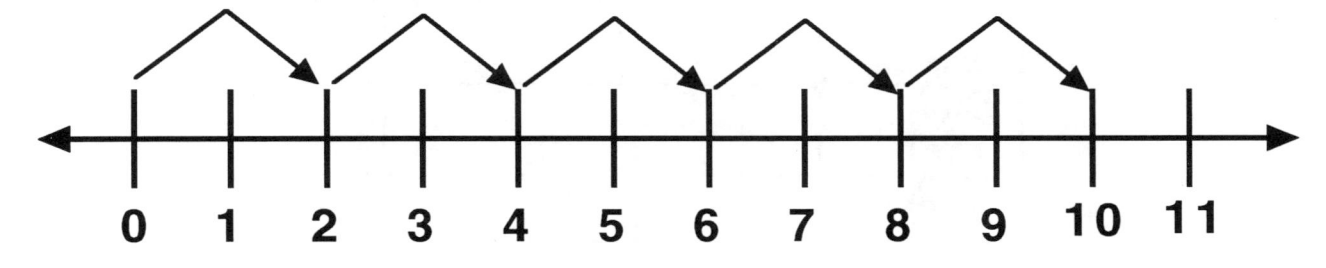

Five moves of two measures on the number line equal ten. A walk-on number line is a good way of illustrating multiplication. Two moves of five on the number line will also equal ten.

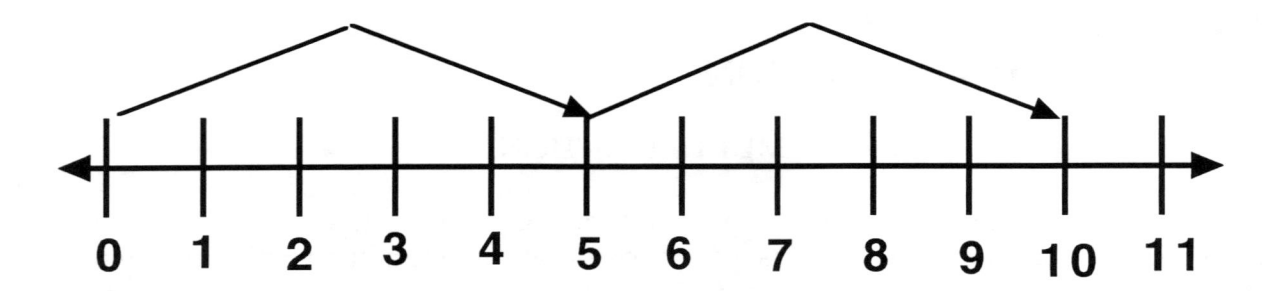

An *array* can also be used to introduce multiplication. An array looks somewhat like a sheet of postage stamps. An array for 3 x 5 would be illustrated as follows:

In the array model, the first number usually stands for the number of sets to be included and the second number indicates how many items are in each set. In some books, the first number represents the number of items per set and the second number represents the number of sets. Which arrangement is used by the teacher is a matter of choice or dictated by the text book, but it should be consistently followed and have the same numbers representing the same idea throughout the lesson.

The *cross products method* (*Cartesian products*) is the fourth arrangement that may be employed when introducing multiplication. Employing this method, 2 x 3 is illustrated below.

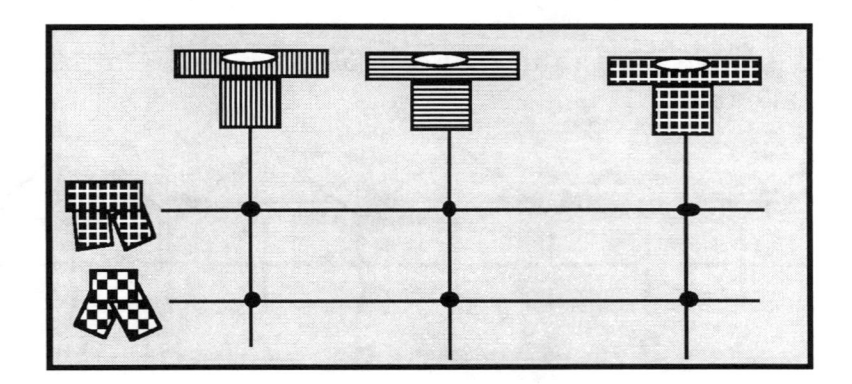

Six is the number of different outfits that could be assembled when there are three T-shirts and two pairs of shorts. Where each line intersects, a different outfit is possible.

Activities for Elementary School Mathematics Page 4.3

DIVISION

Division is usually thought of in two separate ways. One method is called *partitive division* (sometimes called *partition division*), in which the amount is divided into groups with the idea of finding out how many items belong in each group.

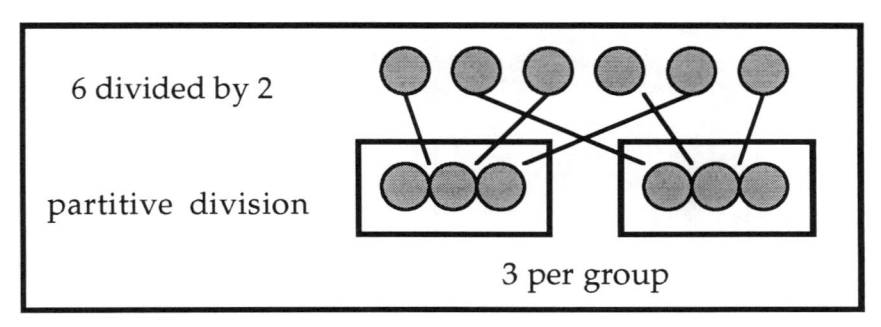

```
6 divided by 2

partitive division

3 per group
```

The other method of division is called *measurement division*, in which the group size is known and the number of groups is the sought-after information.

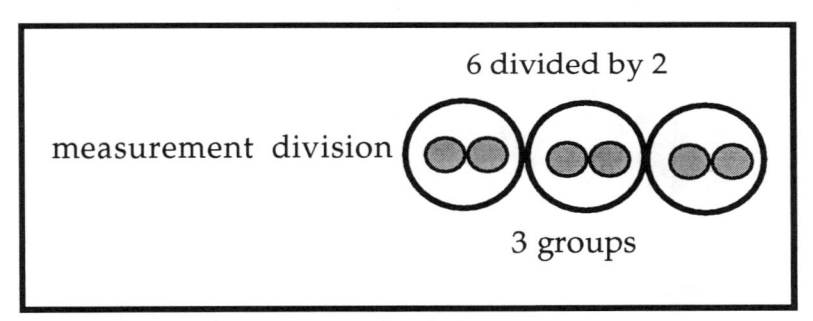

```
                          6 divided by 2
measurement division

                              3 groups
```

Division taught in the classroom is usually illustrated by using partitive division. In the problem below, $76 is to be divided among three people evenly. How much will each person get? Each person gets two ten-dollar bills and there is one ten-dollar bill left over. The one remaining ten-dollar bill is traded in for ten ones and combined with the other six ones to equal sixteen ones. They are divided among the three people, each getting five ones, and there is a one-dollar bill left over.

```
        25 r 1
      ┌───────
   3  │  $7 6
      │    6
      │   ───
      │   1 6
      │   1 5
      │   ───
      │     1
```

Remember that division is the inverse of multiplication. Remind pupils that multiplication and division are related to each other. The activities in this chapter are designed to reinforce this relationship.

Activities for Elementary School Mathematics

WALK-ON NUMBER LINE MULTIPLICATION

Topic: BASIC MULTIPLICATION FACTS

Grade Level: 3-4

Activity Time: 1 class period

Materials Needed:
1. A walk-on number line

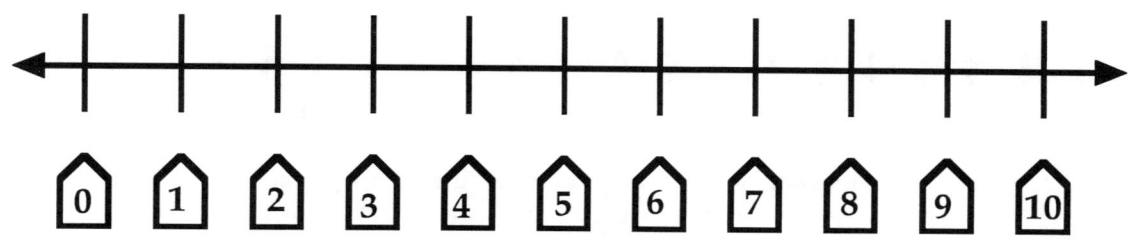

The walk-on number line can be made of heavy plastic, such as an old shower curtain. Affix the number line to the floor with masking tape, using black tape for the center line and cross lines, spaced so pupils can walk on it. The numbers should be placed on the floor next to the number line, with their arrow-like sides pointing to each appropriate intersecting line.

2. Cards (as illustrated above) with a numeral ranging from zero through ten on each of them, made of heavy card stock and laminated for endurance (For larger groups of children, make two sets of cards, one set for each side of the number line. Children can then walk on both sides of the center line.)
3. A list of multiplication facts with products of ten or less for each pupil

Objectives:
As a result of this activity, the learner will:
- understand basic multiplication facts.

Introduction:
1. Review the function of the number line, using addition and subtraction to illustrate.

Major Instructional Sequence:
1. Explain that 4 x 2 means four moves of two cross-lines.
2. Show pupils how to plot 4 x 2 on the number line by taking two steps to the second cross-line and holding up one finger. Then move two more steps to the fourth cross-line and hold up two fingers, two more steps to the sixth cross-line and hold up three fingers, and finally take two more steps to the eighth cross-line and hold up four fingers. Say, "Four times two is eight; I am on the eighth line now."
3. Ask a pupil to plot 2 x 3 by walking on the number line.
4. Ask another pupil to plot a different multiplication fact, such as 3 x 3.

Closure or Evaluation:

A number line of 0 through 10 will not yield enough multiplication facts for each pupil to take a turn. The class may watch as several examples are worked, or a number line of 0 through 20 could also be used. On a sidewalk, tennis court, or on the playground, longer number lines can be drawn so that larger multiplication facts can be explored.

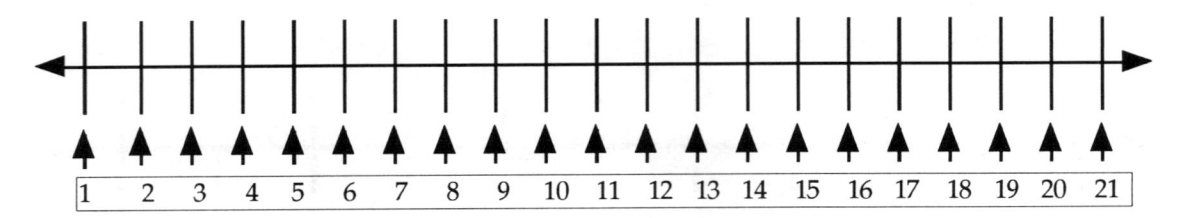

Sidewalk or playground number lines allow pupils
to explore larger multiplication facts.

Activities for Elementary School Mathematics

ROLLO-PRODUCT

Topic: BASIC MULTIPLICATION FACTS

Grade Level: 3-4

Activity Time: 1 class period

Materials Needed:
1. A pair of number cubes with numerals ranging from 1 to 9 on them (one pair for each group of two pupils), with numbers selected by the teacher to reflect the needs of the pupils

2. Paper and pencil for each pupil
3. Optional: a basic facts chart for each pair of pupils playing the game

Objectives:
As a result of this activity, the learner will:
- master basic multiplication facts.

Introduction:
1. Ask pupils to play the game with a partner.
2. Explain to the class that pairs of pupils will work and check each other on their basic multiplication facts.

Major Instructional Sequence:
1. Instruct one pupil in each group to roll a cube to determine who is to start the game. The pupil rolling the highest number starts.
2. The first pupil rolls the cubes and multiplies the resulting two numbers together mentally.

3. If his or her product is correct, that pupil receives one point. If the answer is not correct, the missed combination is recorded on that pupil's paper.

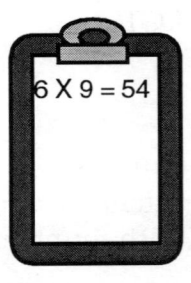

6 X 9 = 54

6 x 9 is recorded if answer is not correctly given by the pupil.

4. The next player rolls the dice, and the same procedure is followed.
5. Circulate among the pupils, checking for understanding, and providing assistance when appropriate.

Closure or Evaluation:
1. All pupils who complete ten correct products win the game. Each player is to be allowed the same number of turns, so the game may end in a tie.
2. Pupils use the missed combinations recorded on the paper for drill.

Activities for Elementary School Mathematics   Page 4.9

MULTIPLICATION WAR

Topic: MULTIPLICATION FACTS

Grade Level: 3-4   **Activity Time:** 1 class period

Materials Needed:
1. Paper and pencil for each pupil
2. A deck consisting of six sets of 3 x 5 index cards, with a number ranging from 0 through 9 on each set (a total of 60 cards for each group of two to four pupils)

| 0 | 1 | 2 | 3 | 4 | 5 | 6 | 7 | 8 | 9 |

Objectives:
As a result of this activity, the learner will:
- recognize larger numbers.
- find the products of two-digit numbers.

Introduction:
1. Divide pupils into groups of two to four.
2. Shuffle the cards and place them, in two stacks of 30 cards each, in the middle of the pupils taking part in the activity.
3. Ask the pupils to draw one card to see who starts the activity. The pupil drawing the highest card in each group starts, and the game progresses to the right.

Major Instructional Sequence:
1. Instruct the first pupil in each group to take the top card from each stack and multiply the values of the cards together.
2. Each pupil will record the product resulting from his or her combination on his or her paper.

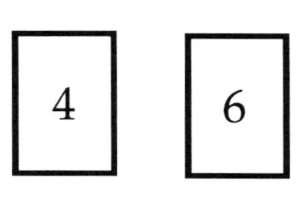

3. The next players in each group will take their turns, following the same instructions. When each pupil in a group has had a turn, round two begins. The game continues until each group has completed a predetermined number of rounds. (The number of rounds possible with 60 cards will vary according to the number of pupils in each group.)
4. After a designated number of rounds, each pupil adds together the products recorded on his or her paper.

Closure or Evaluation:
1. Pupils receive points for the highest total (1 point), second-highest total (2 points), and so on.
2. At the end of the game, the pupil in each group with the lowest number of points is declared champion.

Activities for Elementary School Mathematics   Page 4.11

MARATHON MULTIPLY

Topic: BASIC MULTIPLICATION FACTS

Grade Level: 3-4   **Activity Time:** 1 class period

Materials Needed:
1. Footprints with a multiplication fact on the front of each foot and the product on the back

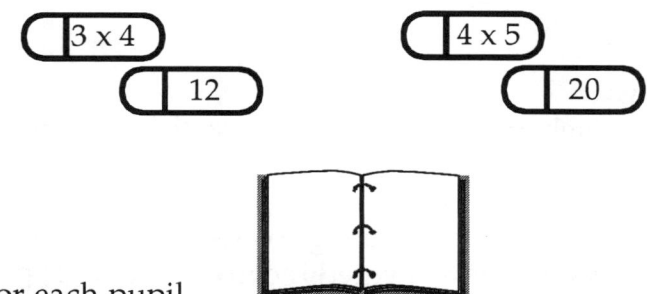

   3 x 4   4 x 5
   12   20

2. A notebook for each pupil

> **NOTE:** *Each pupil needs a notebook to record the multiplication facts they miss on their marathon facts run. Pupils will need to practice any of the facts that they miss; teacher may want to record multiplication facts the pupils missed so they can work with the pupils later.*

Objectives:
As a result of this activity, the learner will:
- become more proficient with basic multiplication facts.

Introduction:
1. Select twenty-six of the footprints. (You may change the footprints as appropriate for pupils with differing levels of ability. Twenty-six footprints are used because marathons are 26.2 miles in length.)
2. Lay the footprints on the floor, problem-side up, in the order desired.
3. Explain the procedure as follows:
   - As you go down the footprints, look at the combination and say the problem and the product aloud.
   - Pick up the footprint and verify your answer by looking on the reverse side.
   - If correct, replace the footprint and proceed to the next footprint and follow the same procedure.
   - If incorrect, record the combination and correct product in your notebook for later study, then replace the footprint and proceed to the next footprint and follow the same procedure.

Major Instructional Sequence:
1. The first pupil goes down the twenty-six step run.
2. Monitor carefully and keep track of the steps missed, making sure the pupil records them in his or her notebook.
3. Pupils continue, in turn, to run the marathon and record missed facts in their notebooks.
4. All pupils go the distance.

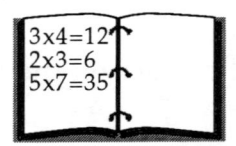

Closure or Evaluation:
Pupils use their notebooks to review and practice missed facts.

Variation:
Let each pupil take a turn, attempting to answer as many multiplication facts as they can. When a fact is answered incorrectly, that pupil sits down and records the multiplication combination missed in his or her notebook. The game continues until all pupils get a chance to participate. If time permits, let pupils go a second time.

Activities for Elementary School Mathematics

Page 4.13

MULTIPLICATION TRACK

Topic: BASIC MULTIPLICATION FACTS

Grade Level: 3-4   **Activity Time:** 1 class period

Materials Needed:
1. Ten sheets of 8.5 x 11 white construction paper, with a numeral between 0 and 9 drawn prominently on each sheet
2. 3 x 5 index cards with a numeral between 2 and 9 on each card
3. A stopwatch (optional)

Objectives:
As a result of this activity, the learner will:
- become more proficient with multiplication facts.

Introduction:
1. Fold each of the ten numbered sheets of construction paper in half so that they will stand upright. Place them in a straight line (or in the shape of a track oval) in random order.

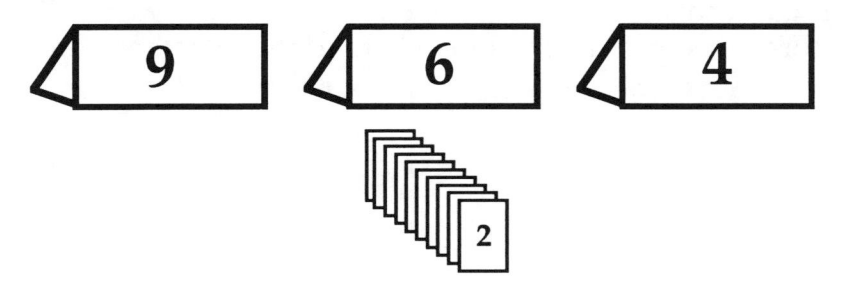

Major Instructional Sequence:
1. The class should be divided into three or four teams, with about 6 players per relay team.
2. Shuffle the 3 x 5 cards and place them face down on a desk.
3. Instruct a pupil from the first team to turn a card face up and read the number on the card.
4. Instruct the pupil to multiply that number times the number on each hurdle and step over the hurdle after answering correctly.
5. The second team sends a member up to turn the next card over and read the number. This pupil takes his or her turn "jumping" the multiplication hurdles.

Closure or Evaluation:
1. Act as scorekeeper for all the teams. Each team member goes through the hurdles and gets one point for each correctly answered fact.
2. If desired, keep track of incorrect responses for later review and practice.
3. The team who scores the most points wins.

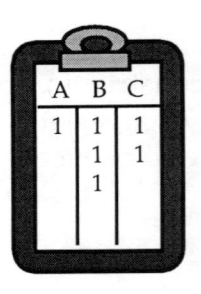

Variation:
A stopwatch could be used to break ties. In addition to your scorekeeping duties, mark down each individual team member's time. If two teams score the same total number of points, add the times of the members of each team. The winning team would be the one with the lowest total time. The stopwatch could also be used to add an extra element of competition. Time individual pupils' responses and add them together at the end of the "track race." The team with the lowest score wins, provided all members of the team answered all the facts correctly.

Activities for Elementary School Mathematics   Page 4.15

MULTIPLICATION MATCH

Topic: MULTIPLICATION FACTS

Grade Level: 3-5   **Activity Time:** 1 class period

Materials Needed:
1. 30 to 50 cards (one-half of a 3 x 5 index card works nicely), each containing a multiplication fact to be practiced
2. 30 to 50 cards with answers to the multiplication facts cards

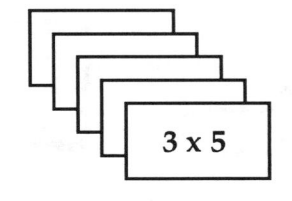

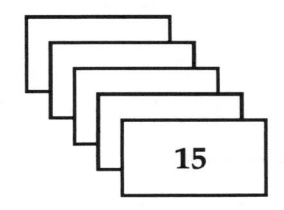

Objectives:
As a result of this activity, the learner will:
- improve his or her knowledge of multiplication facts.

Introduction:
1. Place the answer cards face up upon a desk or table.
2. Place the multiplication facts cards in a stack, face down on the desk or table top.

15	30	35	49	18
10	12	20	42	36
32	16	25	28	21

Major Instructional Sequence:
1. Instruct the first pupil to turn up the top facts card and attempt to match it with one of the answers displayed on the desk or table top.
2. If the pupil makes a match, he or she keeps both cards.
3. Return the facts card to the bottom of the stack if the pupil is unable to make a match.
4. The game continues until all cards are matched or for a specified time.

Closure or Evaluation:
The winner is the pupil with the most matches in his or her possession after all of the facts cards have been played.

MARATHON DIVISION

Topic: BASIC DIVISION FACTS

Grade Level: 3-5   **Activity Time:** 1 class period

Materials Needed:
1. Footprints with a division fact on the front of each foot and the answer on the back
2. A notebook for each pupil

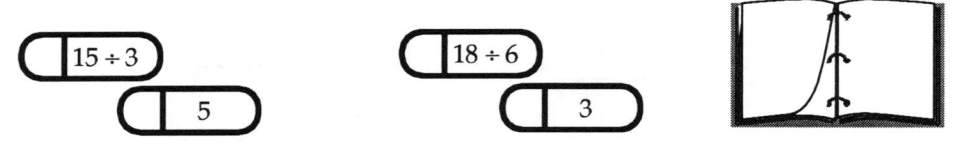

NOTE: *Each pupil needs a notebook to record the division facts they miss on their marathon facts run. Pupils will need to practice any of the facts that they miss; teacher may want to record division facts the pupils missed so they can work with the pupils later.*

Objectives:
As a result of this activity, the learner will:
- become more proficient with basic division facts.

Introduction:
1. Select twenty-six of the footprints. (You may change the footprints as appropriate for pupils with differing levels of ability. Twenty-six footprints are used because marathons are 26.2 miles in length.)
2. Lay the footprints on the floor, problem-side up, in the order desired.
3. Explain the procedure as follows:
   - As you go down the footprints, look at the combination and say the problem and the answer aloud.
   - Pick up the footprint and verify your answer by looking on the reverse side.
   - If correct, replace the footprint and proceed to the next footprint and follow the same procedure.
   - If incorrect, record the problem and correct answer in your notebook for later study, then replace the footprint and proceed to the next footprint and follow the same procedure.

Major Instructional Sequence:
1. The first pupil goes down the twenty-six step run.
2. Monitor carefully and keep track of the steps missed, making sure the pupil records them in his or her notebook.

Activities for Elementary School Mathematics   Page 4.17

   3. Pupils continue, in turn, to run the marathon and record missed facts in their notebooks.

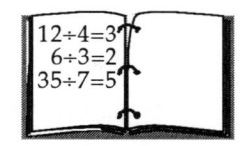

```
12÷4=3
 6÷3=2
35÷7=5
```

   4. All pupils go the distance.

Closure or Evaluation:
    Pupils use their notebooks to review and practice missed facts.

Variation:
    Let each pupil take a turn, attempting to answer as many division facts as they can. When a fact is answered incorrectly, that pupil sits down and records the division combination missed in his or her notebook. The game continues until all pupils get a chance to participate. If time permits, let pupils go a second time.

Page 4.18   Activities for Elementary School Mathematics

BOBCAT!

Topic: DIVISION WITH REMAINDERS

Grade Level: 4-6   **Activity Time:** 1 class period

Materials Needed:
1. Paper and pencil for each player
2. One complete set of problem cards and BOBCAT cards per group of 2 to 6 pupils as follows:
   - Sixty 3 x 5 index cards with a division problem on each card (select problems that result in remainders, for example, 22 ÷ 3). Do *not* include quotients or remainders on the cards. Pupils should be familiar with these types of problems from prior classroom study. Several problem cards per set may result in a quotient with the same remainder.
   - 15 BOBCAT cards with a remainder on each card, as illustrated below. (The remainders should be obtainable by working the problems on the division problem cards.) Several BOBCAT cards per set may have the same remainder.)

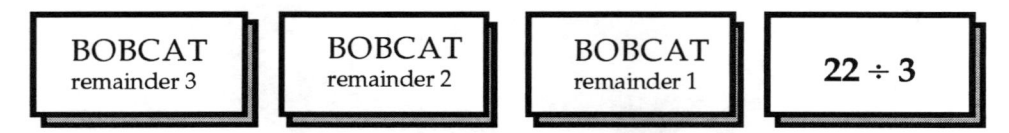

| BOBCAT | BOBCAT | BOBCAT | |
| remainder 3 | remainder 2 | remainder 1 | 22 ÷ 3 |

Objectives:
As a result of this activity, the learner will:
- improve his or her ability to divide problems with remainders.

Introduction:
1. Divide pupils into groups of two to six. Deal the problem cards so that every pupil in a group has the same number of cards to start the game. (There may be cards left over, depending on the number of pupils per group.)
2. Place a set of BOBCAT cards face down in a stack in the middle of each group of players.
3. Instruct each group of pupils to turn one of the BOBCAT cards face up with its remainder showing so that all the pupils in the group can see the remainder on the card.

Major Instructional Sequence:
1. Each pupil in the group searches his/her hand for a problem card resulting in the remainder showing on the upturned BOBCAT card.
2. If the BOBCAT card reads "remainder 3," each pupil in the group having a problem card with a remainder of 3 calls out "BOBCAT!" and lays his/her problem card face up on the playing surface. (Only one problem card per player per round can be played, even though a player might have more than one suitable problem card in his/her hand.)

Activities for Elementary School Mathematics

3. All pupils in the group collaborate in checking the problem cards played to make sure they all result in the remainder on the BOBCAT card. All cards that do match the remainder are discarded and any that do not match are returned to the hand of the player who played the card.
4. The next BOBCAT card in the stack is turned face up, and play proceeds.

Closure or Evaluation:
1. Play continues until one or more players in a group has no problem cards left.
2. The winner is the player in each group who discards all of his/her problem cards first. (Likely, there will be multiple winners who discard all of their cards concurrently, which is perfectly fine.)

Page 4.20   Activities for Elementary School Mathematics

DIVISION MONEY

Topic: DIVISION

Grade Level: 4-6   **Activity Time:** 1 class period

Materials Needed:
1. Chalk and chalkboard
2. Play money ($1, $10, and $100 bills)

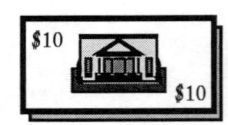

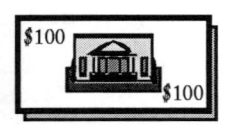

Objectives:
As a result of this activity, the learner will:
- better understand the division algorithm.

Introduction:
1. Write a problem (for example, 764 divided by 3) on the chalkboard.
2. Choose a pupil to act as the banker. (His or her responsibility will be to count out one group of ten $10 bills and one group of ten $1 bills.)
3. Ask for three volunteers to come to the front and assist. Reassure the volunteers that the only thing they have to do is stand in front of the class and put out a hand; no thinking required.

Major Instructional Sequence:
1. Using the example of 764 ÷3, take seven $100 bills and pass them out one at a time to the three pupils. Each pupil gets two $100 bills, and you will have one $100 bill left over. Using the chalkboard, illustrate the first part of the division algorithm to the class.

$$
\begin{array}{r}
2 \\
3\,\overline{\smash{)}\,764} \\
\underline{6} \\
1
\end{array}
$$

Activities for Elementary School Mathematics

2. Take the remaining one $100 bill and go to the banker. Ask him or her to exchange it for ten $10 bills. Combine these with the six $10 bills from the original $764. You should now have sixteen $10 bills.
3. Distribute the $10 dollar bills one at a time until each of the three pupils has five $10 bills and you have one $10 bill left over. Relate the transactions to the algorithm on the chalkboard.

```
      2 5
   ┌──────
 3 │ 7 6 4
     6
     ───
     1 6
     1 5
     ───
       1
```

4. Trade in the one $10 bill to the banker for ten $1 bills and combine with the four $1 bills from the original $764 to get fourteen ones. Distribute the ones to the pupils. Each should get four $1 bills, and you should have two $1 bills left over. Relate the transactions to the division algorithm.

```
      2 5 4                $ 2 5 4  r  $ 2
   ┌────────             ┌──────────
 3 │ 7 6 4             3 │ $ 7 6 4
     6                     6
     ───                   ───
     1 6                   1 6
     1 5                   1 5
     ───                   ───
       1 4                   1 4
       1 2                   1 2
       ───                   ───
         2                     2
```

Closure or Evaluation:
1. Explain that you gave three pupils an equal share of the $764, which was $254 and had a remainder of $2.
2. Divide pupils into groups of three or four and instruct them to work two more problems, using play money and putting it into equal stacks.
3. When all of the groups have done their assigned problems, let each group take a turn at the front of the class, relating their money transactions to their algorithms, by showing their work on the chalkboard.

DIVIDE-A-TREAT

Topic: DIVISION BY ONE-DIGIT DIVISORS

Grade Level: 4-6   **Activity Time:** 1 class period

Materials Needed:
1. A cup for each pupil
2. Paper and pencil for each group of two to five pupils
3. Some kind of treat, such as grapes, cherries, or cookies, in a baggie for each group of two to five pupils (The quantity of treats per bag will depend on the number of members in a group — see note below.)

Objectives:
As a result of this activity, the learner will be able to:
- explain division without remainders.
- explain division with remainders.

Introduction:
1. Before class begins, count out the appropriate number of treats (depending on the size of the groups) and place them in baggies.
2. Divide the class into groups of two to five pupils. Distribute the bags of treats so that each group has an amount that can be divided evenly among the group members.

    NOTE: *A group of two could be given 20 grapes; a group of three could be given 24 cookies; and a group of four could receive 16 cherries.*

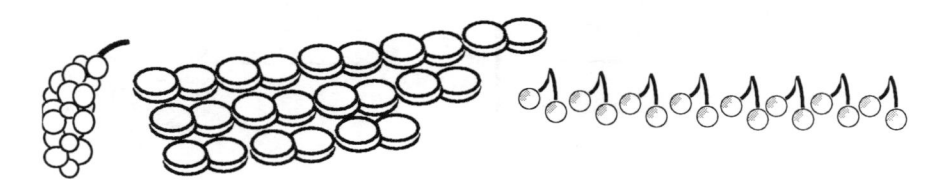

3. Distribute the cups.

Major Instructional Sequence:
1. Instruct each group to figure out how to distribute the treats among its members so that each member gets the same amount.
2. Ask each of the groups to write down the number of treats each group member received and the total number of treats received by the entire group.

Activities for Elementary School Mathematics   Page 4.23

Closure or Evaluation:
1. Ask one pupil from each group to explain how the treats were distributed and then permit the groups to eat their treats.
2. Distribute treats again, this time in amounts that *cannot be* divided evenly among group members. Ask the groups to repeat the process of attempting to divide the treats among group members as evenly as possible (20 ÷ 3 = 6 r. 2)

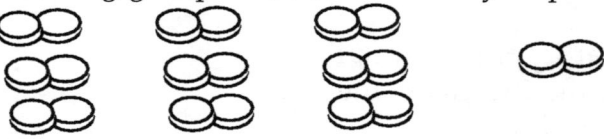

For this example (20 ÷ 3 = 6 r. 2), the group must explain that each of the three members got six cookies and there were two cookies left over.

Thirteen cherries divided among four pupils would mean that each received three cherries and there would be one left over.

As a result of this lesson pupils should realize that division can be visualized as separating items into groups containing equal amounts.

Page 4.24   Activities for Elementary School Mathematics

QUOTIENT BINGO

Topic: BASIC DIVISION SKILLS

Grade Level: 3-6   **Activity Time:** 1-2 class periods

Materials Needed:
1. BINGO game cards, one per pupil
2. Beans to use in the BINGO game
3. Set of division flash cards

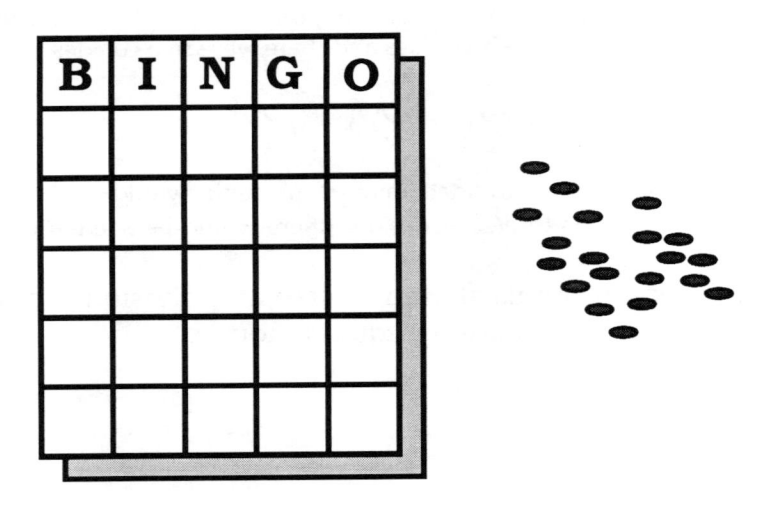

Objectives:
As a result of this activity, the learner will:
- become more familiar with finding quotients.

Introduction:
1. Pass out the BINGO cards and beans to pupils.
2. Instruct pupils to write numerals in the boxes on their cards which might represent answers to division problems to be called out from flash cards. Demonstrate with an example: *"One of the flash cards says 15 ÷ 5, so I know 3 would be a good number to place on the BINGO card. One of the flash cards says 12 ÷ 2, so I know 6 would be a good number to place on the BINGO card."*
3. Display flash cards on the chalktray so that pupils will have appropriate models from which to choose.

Major Instructional Sequence:
1. Shuffle the flash cards and place them in a stack.
2. Select the first flash card and by randomly selecting one of the letters in BINGO, say, "Under the B, 4 ÷ 2." Pause to give pupils time to calculate the quotient and place a bean on the their cards if the answer is there. Then select

the next flash card, randomly select another letter in BINGO, and say, "Under the N, 14 ÷ 2," and so on.

Note: *Write the selections down in a list as they are called, for later checking.*

3. Pupils compute answers mentally and place beans on their BINGO cards if the answer is there.

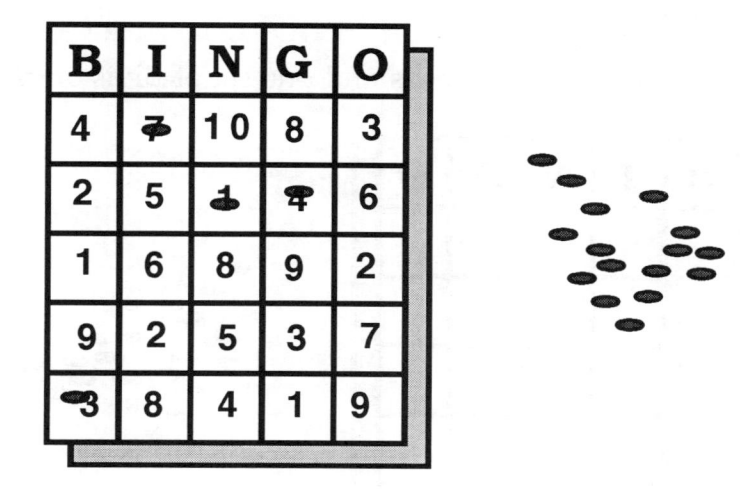

B	I	N	G	O
4	7	10	8	3
2	5	4	7	6
1	6	8	9	2
9	2	5	3	7
3	8	4	1	9

4. Five beans in any direction (down, across, or diagonally) is a BINGO.

Closure or Evaluation:
The game continues until a winner emerges by calling out "BINGO!" and is checked for accuracy from the calling list you have kept.

Variation:
Use the same activity, but as a multiplication exercise. Substitute multiplication flash cards and products for division flash cards and quotients.

COMBINATION BINGO

Topic: BASIC MULTIPLICATION AND DIVISION SKILLS

Grade Level: 3-6   **Activity Time:** 1-2 class periods

Materials Needed:
1. BINGO game cards, one per pupil
2. Beans to use in the BINGO game
3. One set of multiplication flash cards and one set of division flash cards

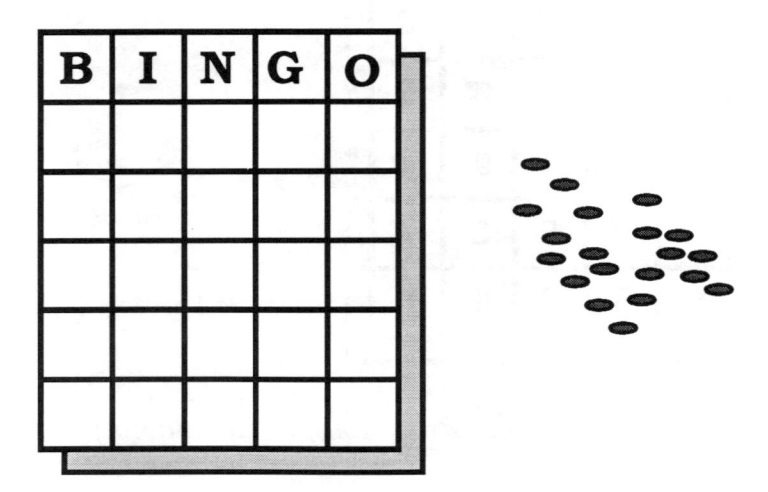

Objectives:
As a result of this activity, the learner will become more familiar with:
- finding products.
- finding quotients.

Introduction:
1. Pass out the BINGO cards and beans to pupils.
2. Instruct pupils to write multiplication and division fact combinations in the boxes on their cards which might represent the combinations for products and quotients to be called out from flash cards. Demonstrate with an example: "The answer on one division flash cards says 3, so I know 15 ÷ 5 or 12 ÷ 4 or 6 ÷ 2 would be good combinations to place on the BINGO card. The answer on one multiplication flash cards says 14, so I know 7 x 2 would be a good combination to place on the BINGO card."
3. Display flash cards on the chalktray (showing the answer side) so that pupils will have appropriate models from which to choose. Suggest to pupils that they put about half multiplication and half division combinations on their cards.

Activities for Elementary School Mathematics   Page 4.27

Major Instructional Sequence:
1. Count out ten division cards and ten multiplication cards, shuffle them, and lay them in a stack.
2. Select the first flash card, look at the answer side, and by randomly selecting one of the letters in BINGO, say, "Under the B, 7." Pause to give pupils time to think and write, then select the next flash card, randomly select another letter in BINGO, and say, "Under the N, 14," and so on.
3. When you have exhausted the first stack of cards without a "BINGO," count out ten more division and ten more multiplication cards, shuffle them, make a new stack, and continue.

Note: *Write the selections down in a list as they are called, for later checking.*

4. Pupils compute answers mentally and place beans on their BINGO cards if the answer is there.

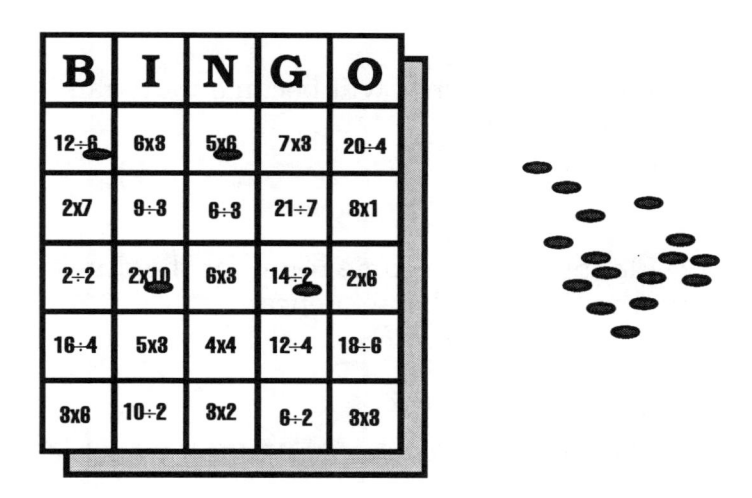

B	I	N	G	O
12÷6	6x3	5x6	7x3	20÷4
2x7	9÷3	6÷3	21÷7	8x1
2÷2	2x10	6x3	14÷2	2x6
16÷4	5x3	4x4	12÷4	18÷6
3x6	10÷2	3x2	6÷2	3x3

5. Five beans in any direction (down, across, or diagonally) is a BINGO.

Closure or Evaluation:
The game continues until a winner emerges by calling out "BINGO!" and is checked for accuracy from the calling list you have kept.

Activities for Elementary School Mathematics

DOMINO MULTIPLICATION

Topic: BASIC MULTIPLICATION FACTS

Grade Level: 3-6

Activity Time: 1 class period

Materials Needed:
1. One set of dominoes for the class

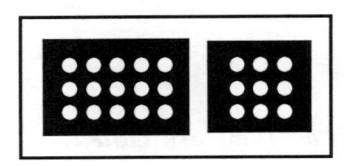

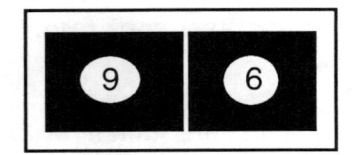

   The teacher could make a special domino set to cover higher multiplication facts if desired.

2. Paper and pencil for each pupil

Objectives:
As a result of this activity, the learner will become more familiar with:
- number/numeral concepts.
- multiplication facts.

Introduction:
1. Put desks in a circle (or establish some route for the dominoes to travel from one person to another in a regular sequence).
2. Furnish each pupil with one domino.

Major Instructional Sequence:
1. Tell the pupils to turn the dominoes face up on their desks.

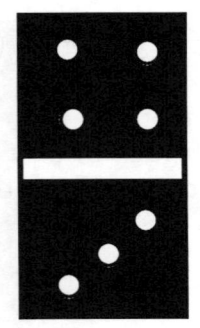

   4 x 6 = 24

   3 x 4 = 12

2. Explain to the pupils that they are to write a multiplication problem and answer (number sentence) for each domino they receive.
3. After they have written their problems, give a signal for pupils to pass their dominoes along in the predetermined direction.
4. Upon receiving a new domino, each pupil creates new problems (number

Activities for Elementary School Mathematics

sentences). A predetermined number of exchanges can be established for the game, or the game can continue until all of the dominoes have been passed around the class and returned to the point where they started.

Closure or Evaluation:
1. Have pupils volunteer to read their number sentences aloud, and the rest of the class gives a "thumbs-up" for correct number sentences and a "thumbs-down" for incorrect number sentences. (Monitor this activity carefully. Try to see that everyone has an opportunity to participate without anyone being placed in an embarrassing situation.)
2. Ask pupils to circle any incorrect number sentences on their papers during this process.
3. Collect the pupils' papers so that you can check them for individual understanding.

Note: *If pupils find and mark the first number sentence (when it is read aloud) on their papers, the rest of the number sentences should appear in order, making the checking process easier.*

DIVISION RUMMY

Topic: DIVISION SKILLS

Grade Level: 3-6   **Activity Time:** 1 class period

Materials Needed:
1. Paper and pencil for each pupil
2. Identical decks of 3 x 5 index cards, each deck consisting of 50 cards, per group (Each deck has 25 division problem cards and 25 quotient [answer] cards, and it is acceptable for several problem cards to have the same quotient.)

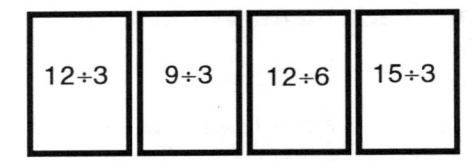

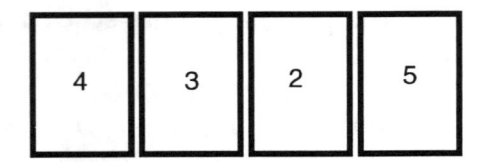

Objectives:
As a result of this activity, the learner will:
- find the quotient of a division problem.
- practice division skills.

Prior to beginning the activity, explain that the object of the game is to be the person that collects the most matches between quotient cards and problem cards and has no other cards left. (Demonstrate some possible matches.)

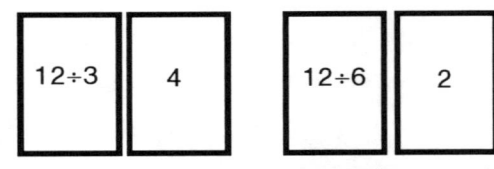

The first two cards match because 12 ÷ 3 = 4. The second two cards match because 12 ÷ 6 = 2.

Introduction:
1. Divide pupils into groups of four.
2. Give each group a deck of cards.
3. Ask one person at each table to be the dealer, shuffle the cards, and deal ten cards to each person in the group. (Put left-over cards in a "draw" pile.)
4. Play begins with the person on the dealer's right and progresses around the group to the right.

Major Instructional Sequence:
1. Instruct the dealer in each group to start the game by turning up the top card in the "draw" pile.

Activities for Elementary School Mathematics

2. The next pupil (to the right) plays a card which either has the quotient (answer) for the card on the table or a problem which would yield the quotient for the card on the table, then collects both cards and plays a new card.
3. If the player has no card that will "play," he or she picks up a card from the "draw" pile, and plays it if it will "play;" if not, the card is added to the player's hand and play continues to the right.
4. Play continues until a player has no more cards to play and has the most sets of matches. (If a pupil has no more cards to play but doesn't have the most matches, and there are no more cards in the "draw" pile, that player "passes" until the game is completed.)
5. Pupils may use their pencils and paper for computation, if necessary, but must work reasonably fast and cannot "bog down" the game.

Closure or Evaluation:
The pupil in each group who has the most wins after five rounds of play is declared the group champion. A grand champion can be determined by the group champion with the most sets of matches (ties are likely, and co-grand champions can be named).

Variation:
Each group champion can compete in a "playoff" round with the other group champions to determine a grand champion.

SUBTRACTIVE DIVISION

Topic: DIVISION SKILLS

Grade Level: 5-6

Activity Time: 1 class period

Materials Needed:
1. Paper and pencil for each pupil
2. Six plastic zip-top baggies, each filled with 100 beans
3. Chalk board and chalk

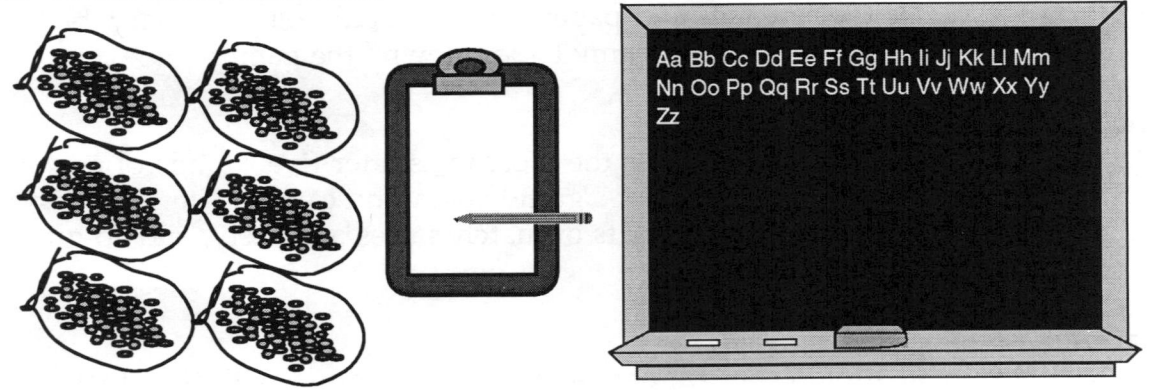

Objectives:
As a result of this activity, the learner will:
- use concrete manipulatives to perform subtractive division.
- perform subtractive division with pencil and paper.

Prior to beginning the activity, demonstrate on the chalkboard how subtractive division functions (see below). Explain to pupils that they will work in groups and use various amounts of beans to work-out subtractive division problems.

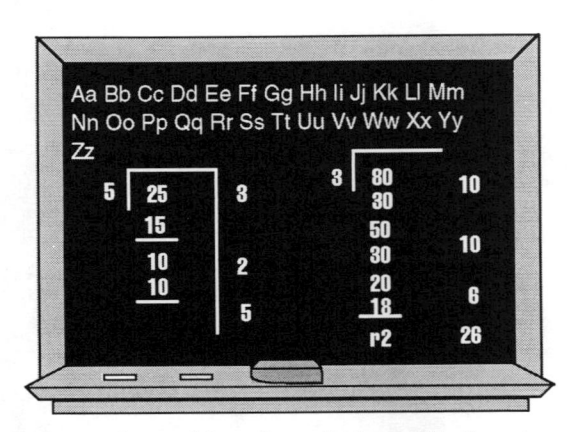

The example on the left shows first taking three 5s from 25, then two, revealing five 5s in 25. The example on the right shows taking ten 3s, then another ten 3s, then six 3s revealing twenty-six 3s in 80, with a remainder of 2.

Activities for Elementary School Mathematics   Page 4.33

Introduction:
1. Divide pupils into six groups.
2. Give each group a bag of beans.
3. Place a division problem on the chalkboard within the capabilities of the children in the class.

Major Instructional Sequence:
1. Instruct each group to work collaboratively to put the correct number of their beans into a cluster which represents the dividend. (For example, if the problem you've placed on the board is 30 ÷ 6, the groups would place 30 beans in their clusters.)
2. Lead the groups in a discussion which reveals that the "divisor" is 6 for the problem on the chalkboard.
3. Instruct each group to work collaboratively to find how many sets of 6 beans can be taken from the cluster of 30 beans. (Circulate, check for understanding, and give assistance as needed, making sure that all groups subtracted five sets of six beans from the cluster.)
4. Ask pupils to collaborate in their groups to see if there are other ways to subtract the beans faster and still get the correct answer. For example, three sets of six beans plus two sets of six beans, or two sets plus two sets plus one set, and the like. Have each group write their various combinations on paper, using the form you used on the board originally.
5. Ask groups to volunteer to demonstrate one of their combinations on the chalkboard. Have groups continue to volunteer to demonstrate one combination until all of the possibilities discovered by all groups have been demonstrated. (On the chance that a group demonstrates an incorrect combination, have a class discussion on why it is incorrect.)

Closure or Evaluation:
1. Place another division problem on the chalkboard and follow the same sequence.
2. Continue for as long as the class period permits.

Page 4.34   Activities for Elementary School Mathematics

REAL ESTATE BARON

Topic: COMPUTATIONAL SKILLS, INCLUDING DIVISION

Grade Level: 5-6   **Activity Time:** 1 class period

Materials Needed:
1. Paper and pencil for each pupil
2. Six diagrams like the one below on 8.5 x 11 paper
3. Chalk board with same diagram

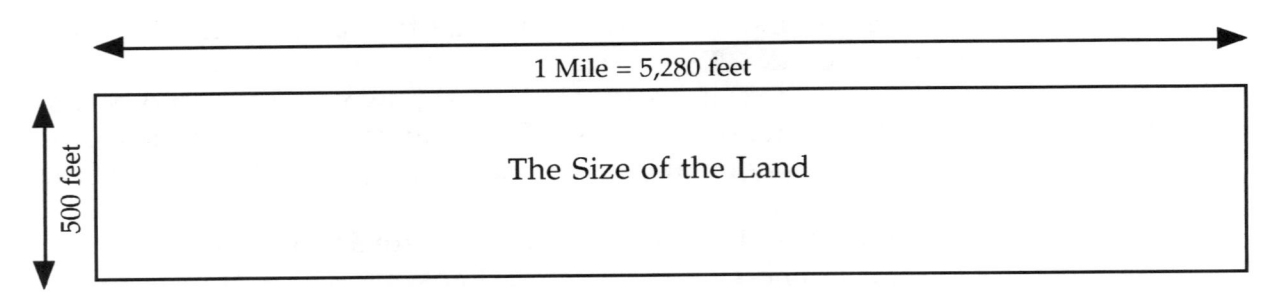

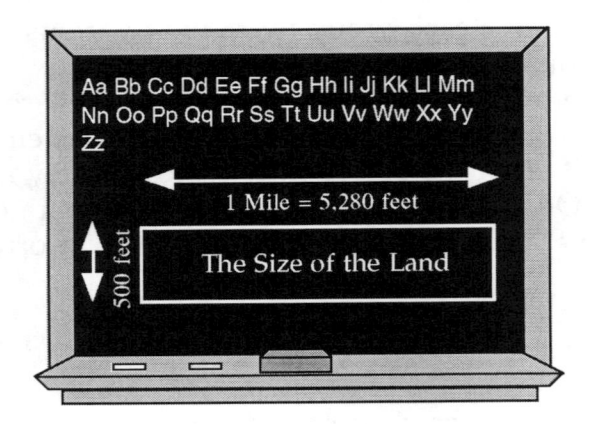

<--------------------- 1 Mile = 5,280 feet --------------------->

↕ 500 feet

The Size of the Land

Objectives:
As a result of this activity, the learner will:
- use higher-order thinking skills in a collaborative group.
- select appropriate types of computations to answer questions.
- solve addition, subtraction, multiplication, and division problems.

Introduction:
1. Divide pupils into six groups.
2. Give each group a sheet showing a diagram of the land size.
3. Using the chalkboard illustration, explain the size and shape of the parcel of land to be considered in this activity.
3. Make sure pupils relate the drawing on their 8.5 x 11 sheets to the drawing on the chalkboard.

Activities for Elementary School Mathematics

Major Instructional Sequence:
1. Instruct each group to work collaboratively to decide how much money their group will make if the land is divided into:
   - 1000 feet wide by 500 feet deep lots selling for $8000 each
   - 500 feet wide by 500 feet deep lots selling for $4800 each
   - 200 feet wide by 500 feet deep lots selling for $3000 each
   - 100 feet wide by 500 feet deep lots selling for $1850 each
2. Tell the groups that if they have land left over in any of their calculations, they must decide how the left-over land is to be used.
3. Ask the groups to decide which of the ways of dividing the land would be the most profitable for the group (1000 x 500, 500 x 500, 200 x 500, or 100 x 500).
4. Circulate among the groups, monitoring carefully, checking for understanding, and giving assistance when appropriate.

Closure or Evaluation:
1. Let each group take turns reporting on their findings.
2. Have each group also explain how they decided to use the left-over land.

LET THE LIGHT SHINE ON ME

Topic: BASIC DIVISION FACTS

Grade Level: 4-6 **Activity Time:** 1 class period

Materials Needed:
1. Flashlight
2. Division facts sheet (8.5 x 11) for each pupil

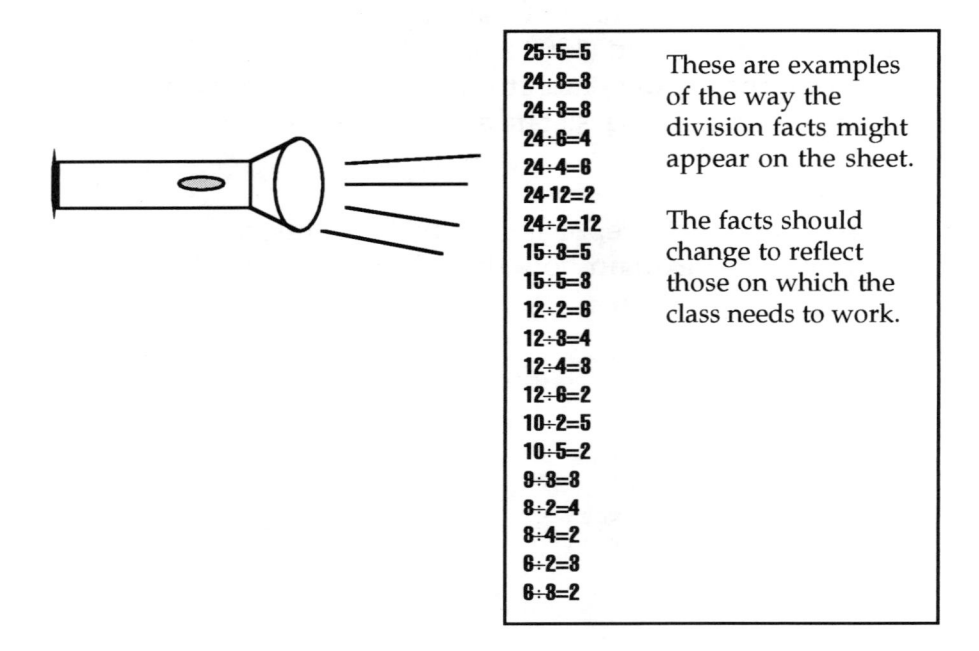

25÷5=5
24÷8=3
24÷3=8
24÷6=4
24÷4=6
24-12=2
24÷2=12
15÷3=5
15÷5=3
12÷2=6
12÷3=4
12÷4=3
12÷6=2
10÷2=5
10÷5=2
9÷3=3
8÷2=4
8÷4=2
6÷2=3
6÷3=2

These are examples of the way the division facts might appear on the sheet.

The facts should change to reflect those on which the class needs to work.

Note: *This activity is more fun if you can dim the lights, draw the window shades, and the like.*

Objectives:
As a result of this activity, the learner will:
- become more proficient with basic division facts.

Introduction:
1. Tell the class that they are going to have a chance to "shine" in division.
2. Give each pupil a sheet with the division facts.
3. Tell pupils that you are going to call out a fact, such as "12 ÷ 3," and they are to raise their hands if they know the answer. (They may look on their division facts sheets, if necessary.)
4. Tell them you will select someone with his/her hand up, and shine the light on that person to give the answer.
5. If the light shines on a pupil, he or she should give the entire fact, with the answer. For example, "12 ÷ 3 = 4."

Activities for Elementary School Mathematics

Major Instructional Sequence:
1. Call out a basic division fact.
2. Select someone who has his/her hand up.
3. Shine the light on that pupil.
4. If the response is correct, pass the flashlight to that pupil and let him/her select the next division fact and follow the same procedure.

Closure or Evaluation:
Continue until the class period is over or until the pupils have practiced the division facts extensively.

DIVISIBLE SHUFFLE

Topic: BASIC DIVISION FACTS

Grade Level: 4-6 **Activity Time:** 1 class period

Materials Needed:
1. Paper and pencil for each pupil
2. Stack of 3 x 5 index cards with numerals 10 through 99 written on them

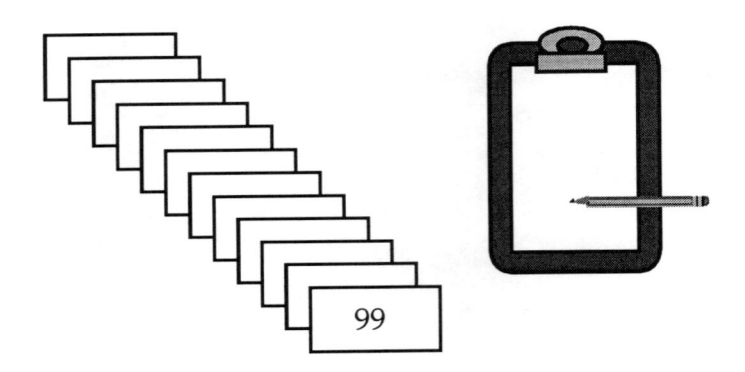

Objectives:
As a result of this activity, the learner will:
- divide two-digit numbers by one-digit divisors.

Introduction:
1. Tell the class to make three columns on their papers.
2. Ask pupils to write a single-digit numeral at the top of each column. (Pupils decide on their own which three numbers to write, as long as they write numbers which are less than ten, but not zero.)

3	5	9

3. Tell pupils that when you call out a two-digit number, like "36" for example, they are to decide which of their numerals will divide into that number without a remainder. (Remind pupils that it's possible that one, two, or all three of their numbers could work, or it might be that none of their numbers

will work. Say, "As an example, if your numerals were 3, 6, and 9, then all three would divide into 36 without a remainder.")

Major Instructional Sequence:
1. Shuffle the index cards.
2. Select the first card and read the two-digit numeral to the class.
3. Tell pupils to write the two-digit number you called out in each column where the number at the top will divide into it without a remainder.
4. Circulate among the pupils, monitoring carefully, checking for understanding, and providing assistance when appropriate.

Closure or Evaluation:
1. After the activity has been concluded, collect the papers so that you can check to see if all pupils have a good understanding of dividing two-digit numbers by one-digit divisors.
2. Schedule additional similar activities, especially for those who may need additional practice.

Variation:
Schedule additional similar activities where you select larger, more complex numerals for the activity.

Page 4.40   Activities for Elementary School Mathematics

CROSS OFF

Topic: BASIC DIVISION FACTS

Grade Level: 4-6   **Activity Time:** 1 class period

Materials Needed:
1. Paper and pencil, a blank grid sheet (may be laminated for multiple use), and a multiplication chart for each pupil
2. Two sets of 3 x 5 index cards with numerals 1 through 9 written on them

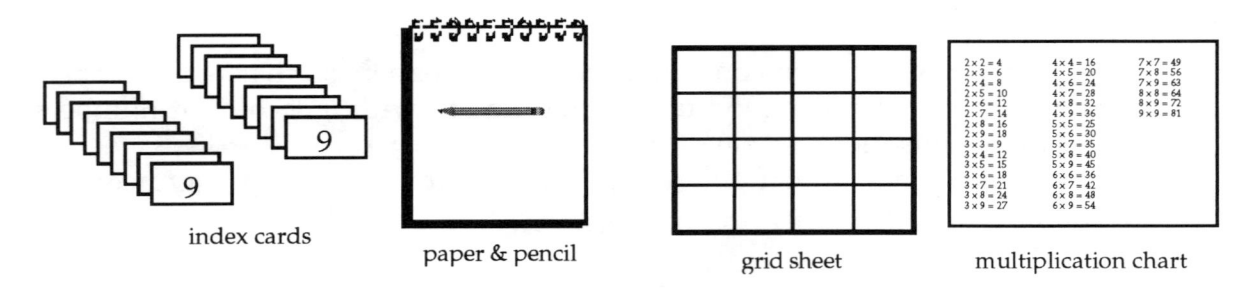

index cards   paper & pencil   grid sheet   multiplication chart

Objectives:
As a result of this activity, the learner will:
- recognize composite numbers.
- divide composite numbers and get an answer with no remainder.

Introduction:
1. Provide each pupil with a 4 x 4 grid on a sheet of paper, like the one below, and a multiplication chart.
2. Instruct pupils to write a composite number in each space on the chart. (Explain that a composite number is one which is a multiple of other numbers. Use the chart to demonstrate several examples, such as 9 x 8 = 72, a composite number, 6 x 7 = 42, a composite number, and the like. Tell pupils to select the composite numbers for their grids from the ones on the multiplication chart.)

2 x 2 = 4	4 x 4 = 16	7 x 7 = 49
2 x 3 = 6	4 x 5 = 20	7 x 8 = 56
2 x 4 = 8	4 x 6 = 24	7 x 9 = 63
2 x 5 = 10	4 x 7 = 28	8 x 8 = 64
2 x 6 = 12	4 x 8 = 32	8 x 9 = 72
2 x 7 = 14	4 x 9 = 36	9 x 9 = 81
2 x 8 = 16	5 x 5 = 25	
2 x 9 = 18	5 x 6 = 30	
3 x 3 = 9	5 x 7 = 35	
3 x 4 = 12	5 x 8 = 40	
3 x 5 = 15	5 x 9 = 45	
3 x 6 = 18	6 x 6 = 36	
3 x 7 = 21	6 x 7 = 42	
3 x 8 = 24	6 x 8 = 48	
3 x 9 = 27	6 x 9 = 54	

Activities for Elementary School Mathematics

Major Instructional Sequence:
1. Shuffle the index cards.
2. Select the first card and read the numeral to the class.
3. Tell pupils to examine their grid sheets to see if they have a composite number which is divisible by the number called. If so, they may cross it off their grid sheets. (Only one composite number may be crossed off each time, even though there may be other composite numbers on the grids which are divisible by the number called.)
4. Monitor the activity carefully by circulating among the pupils, checking for understanding, and providing assistance when appropriate.

Closure or Evaluation:
1. Play continues until a player has crossed off four in a row, a column, or diagonally, and is declared the winner.
2. Play additional rounds as time permits. Pupils may create new grid sheets for new rounds or continue to play with their original grid sheets.

Activities for Elementary School Mathematics   Page 5.1

CHAPTER FIVE
FRACTIONS: COMMON • DECIMAL

Fractions are often introduced into early grades before children are developmentally ready for the concept. Teachers need to be aware that premature introduction of fractions is not a good idea.   As teachers, when we introduce fractions, we must be careful that we do not pass along misconceptions to the pupils. Two common misconceptions are often inadvertently communicated to children. The first is that anything divided into two parts makes the parts halves.

"Wendi, divide the candy bar into halves and give Kristi half."

| Wendi's | Kristi's |

Kristi may be too young to understand fractions, but she knows that she did not get half!

It isn't uncommon to hear people say, "I'll take the smaller half of the candy bar when you split it." There is no such animal as "the smaller half." Half implies, or *should* imply, two pieces of exactly the same size.

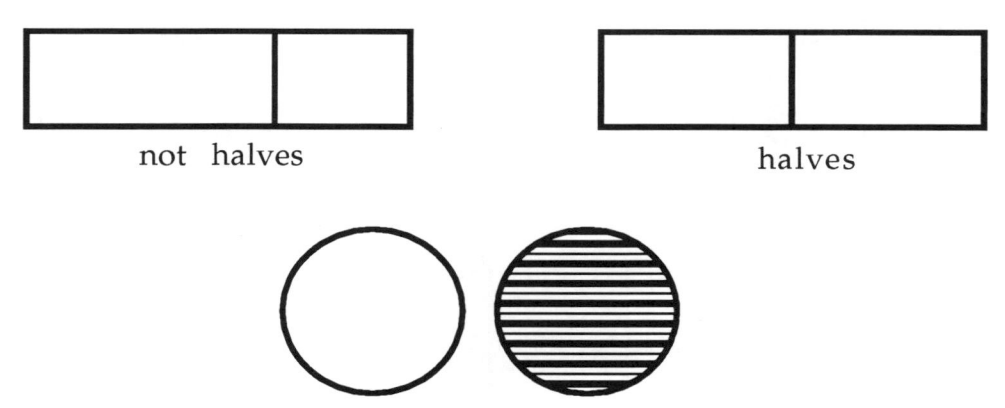

not halves   halves

half (1/2) of the circles have stripes

Another misconception we accidentally pass along to pupils is that 1/2 is always larger than 1/4. This is not true on every occasion, as demonstrated by the figure below.

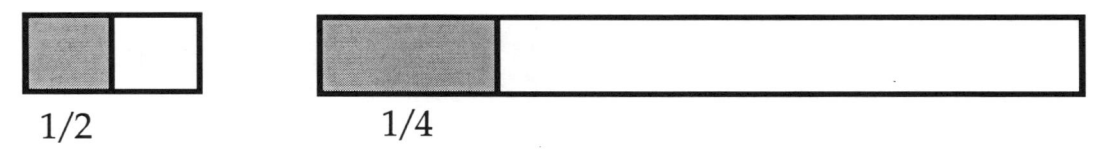

1/2   1/4

In this example, the 1/4 is much larger than the 1/2. In like fashion, 1/4 of $100

Page 5.2   Activities for Elementary School Mathematics

is $25, and 1/2 of $10 is $5, so 1/2 is not always the larger (or, by extension, more desirable) of the two fractions being compared. However, 1/2 of something is always bigger than 1/4 of something, *if it comes from a unit of equal size.*

    Paper folding is an excellent way to introduce fractions. One of the activities in this section takes units of the same size, folds them into fractional parts, and compares the fractions to one another.

           The fraction 1/2   The fraction 1/4

    Other activities in this chapter compare fractions, decimals, and percents in a way that gives pupils hands-on experiences with these concepts.

Activities for Elementary School Mathematics   Page 5.3

FRACTION INTRODUCTION

Topic: FRACTIONS

Grade Level: 2   **Activity Time:** 1 class period

Materials Needed:
1. One set of geometric regions for every two children (Each set should contain one of each: square, circle, oval, triangle, and rectangle.)
2. One pair of scissors for every two children

Objectives:
As a result of this activity, the learner will:
- become familiar with basic fractions.
- be able to discuss what such fractions actually are.

Introduction:
1. Divide pupils into groups of two.
2. Distribute one set of regions to each group.
3. Explain to pupils that when they fold their regions in half, they will have two pieces which are exactly the same size. Further explain that when they fold the same prefolded region in half again, they will have four pieces of the same size. (Demonstrate the process.)
4. Next, distribute scissors so that each group of two pupils has a pair.

Major Instructional Sequence:
1. Ask each group to work collaboratively and cooperatively to fold their regions so that they have pieces that are the same size.

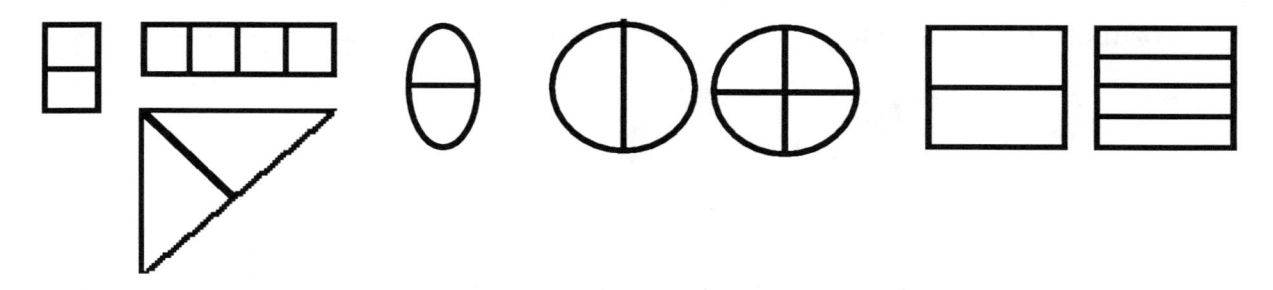

2. Next, instruct pupils to use their scissors and cut the regions along the folds they made.

Closure or Evaluation:
1. Encourage class discussion about what the pupils did with their regions.
2. Ask pupils to explain to the class what they have done and show their results.

Page 5.4   Activities for Elementary School Mathematics

FRACTION FOLD

Topic: FRACTIONS

Grade Level: 2-4   **Activity Time:** 1 class period

Materials Needed:
1. Approximately eight 2 x 11 paper strips per pupil (a good use for scrap paper)
2. Pencils or washable marking pens for each pupil

Objectives:
As a result of this activity, the learner will be able to:
- visualize fractions.
- visualize the difference between fractions.
- determine congruent fractions.

Introduction:
1. The class will be looking at fractions such as 1/2, 1/3, 1/4, 1/5, 2/3, and 3/4 (whatever fractions are to be studied in class).
2. Explain to the pupils that by folding paper, they will actually be able to "see" the fractions, compare different fractions, and find equivalent fractions.

Major Instructional Sequence:
1. Distribute the paper strips to the class. (The amount per pupil will vary, depending on how many fractions are to be "folded.")
2. Instruct each pupil to fold the first piece of paper in the middle, as indicated in the illustration below. Explain that halves will be the result of this fold.
3. Ask the pupils to trace the crease resulting from the fold with their pencils or marking pens. Then instruct the class to shade 1/2 of the region.

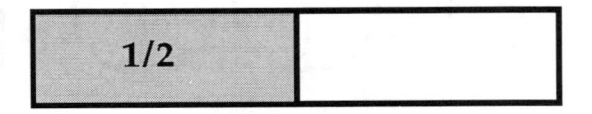

1/2	

4. Show the class how to use the folding process to find 1/3 (Make an **"S"** out of a strip of paper, then flatten it to make 1/3). Instruct the pupils to shade one of the sections.

1/3		

Activities for Elementary School Mathematics   Page 5.5

5. Make 1/4 by folding a paper in half, then in half again.

   | 1/4 | | | |

6. For 1/5, make two-and-a-half loops, then flatten them. (A good way to illustrate this procedure is to use a pupil with a thin wrist as a model.)

   | 1/5 | | | | |

7. The fraction 1/6 is made by following the procedure to create 1/3 and then folding it in half.

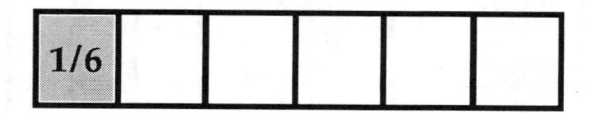

   | 1/6 | | | | | |

8. The fraction 1/8 is 1/2 folded in half, then in half again (three folds).

   | 1/8 | | | | | | | |

Closure or Evaluation:
1. Each time pupils fold the paper strips to visualize a fraction, instruct them to shade one section and label the fractional part by its name, for example, as 1/3, or 1/4. Encourage pupils to make comparisons, such as "1/2 is larger than 1/3," "1/3 is larger than 1/4," "1/2 is larger than 1/4," and so on, because all of the strips of paper are the same size.
2. Based on these comparisons, ask the class questions, such as "How many fourths would it take to make 1/2?" (two).

   | 1/2 | |
   | 1/4 | | | |

3. Other comparisons can be made to visualize fractions and equivalents.

Page 5.6   Activities for Elementary School Mathematics

BINGO FRACTIONS

Topic: FRACTIONS

Grade Level: 2-4   **Activity Time:** 1 class period

Materials Needed:
1. Teacher-made fraction BINGO game cards for each pupil (A sample card appears below.)

To speed up the game, the cards can have more than one of a particular fraction on them. The pupils can cover only one fraction per call, or for a shorter game they can cover all the called fractions at the teacher's discretion.

Activities for Elementary School Mathematics   Page 5.7

2. 3 x 5 index cards with fractions on them (Sample fraction cards are shown below. When creating these cards, select fractions that the class has previously discussed, and make sure that the fraction "pictures" on the BINGO cards match the fractions on the index cards.)
3. Three sets of "calling cards" with the letters **B I N G O** on them

Fraction Cards:	1/2	3/4	1/5	3/8	5/6
Calling Cards:	B	I	N	G	O

Objectives:
As a result of this activity, the learner will be able to:
- identify fractional parts of a region.
- identify fractional parts of a set.

Introduction:
1. After class discussion about fractions as parts of a region or parts of a set of objects, distribute the BINGO game cards to the pupils.
2. Act as the caller for the first game. After the pupils learn the game, a pupil caller (perhaps the winner of the previous game) could be utilized.

Major Instructional Sequence:
1. Make sure pupils know the rules: Tell them they can cover either *one* of the called fractions or *more than one*, depending on the time limits you set for the game.
2. Explain that the shaded portion represents the fractional part of a set.
   For example:

   ○●●● = 3/4

3. Separately shuffle both the fraction cards and the BINGO calling cards and place each stack face down on the table or desk. Pick up the first BINGO calling card from its face-down stack. Also select one of the fraction cards, and read the resulting combination of letter and fraction aloud to the class. For example, if "I-1/3" was called, pupils would look on their cards and cover the appropriate square if they had it.
4. As you call out each fraction, write it in the appropriate column of a record sheet (to facilitate verification of BINGOs). See illustration on the next page.

B	I	N	G	O
3/8	1/3	3/4		1/3
	2/3			

Closure or Evaluation:
   The winner is determined as in regular BINGO (5 across, down, or diagonally).

Activities for Elementary School Mathematics

Page 5.9

FRACTIONAL MATCH

Topic: FRACTIONS

Grade Level: 3-5

Activity Time: 1 class period

Materials Needed:
1. 3 x 5 index cards, each listing a fraction being studied in class, approximately 40 fraction cards per every 3 pupils (Duplicate cards may be needed, depending upon the number of pupils in the class and the number of fractions being studied.)

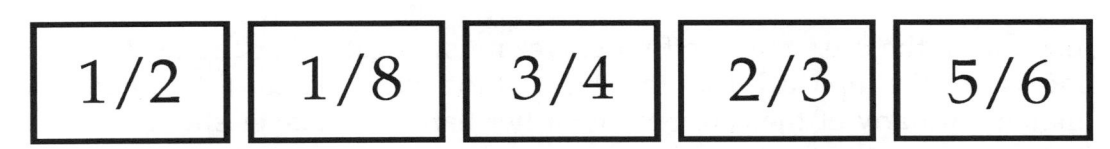

| 1/2 | 1/8 | 3/4 | 2/3 | 5/6 |

2. Another set of index cards with fractional parts or sets shaded to represent each fraction card created

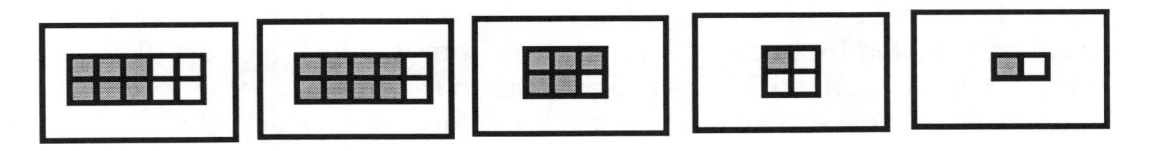

(If equivalent fractions are being studied the following cards could be included in each set for 1/2.)

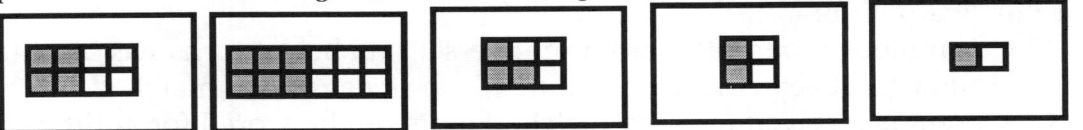

Objectives:
As a result of this activity, the learner will be able to:
- identify fractions by regions or sets.
- match fractions with regions.
- identify equivalent fractions.

Introduction:
1. Before class, make sure that you have created enough cards so that each group will have a set of fractions and fractional regions or sets to match. (If you are using three pupils per group, you will need a deck of about forty cards.)
2. Separate pupils into groups, choose a dealer for each group, and provide each dealer with a deck of cards.

Page 5.10   Activities for Elementary School Mathematics

Major Instructional Sequence:
1. Instruct the dealers to shuffle the cards and distribute seven cards to each player in his or her group.
2. The remaining cards are placed face down in a stack in the middle of the group.
3. One card is turned face up next to the stack of cards.
4. The player to the right of the dealer begins the game.

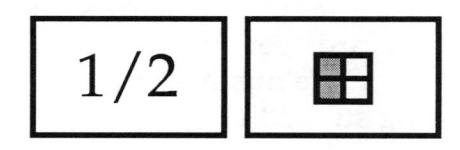

| 1/2 | ⊞ |

5. Explain to the class that the first player must select either the card which is face up or the top card from the stack. If the upturned card does not create a match with any of the cards in his or her hand, the player should choose the top card from the stack. If the player is able to make a match between the upturned card and the cards in his or her hand, the matching cards should be laid face up in close proximity to that player.
6. Whether or not a match has been made, a card must be discarded before the next player takes a turn. The discards will be placed face up in the stack which has been created by turning the first card from the deck face up. The next player takes a turn, with play progressing to the right.
7. After all cards have been selected from the face-down stack, the "discard" pile is shuffled and placed face down as the new "draw" stack.

Closure or Evaluation:
1. Play continues around the group until a set number of turns has been taken, a time limit has been reached, or until all cards have been matched.
2. One point is received for each match. Points can be scored for additional cards placed on initial pairs, in the case of duplicate cards and/or equivalent fractions.

Activities for Elementary School Mathematics

EQUIVALENT BINGO

Topic: EQUIVALENT FRACTIONS

Grade Level: 4-6   **Activity Time:** 1 class period

Materials Needed:
1. Set of 50 fraction cards that are not in simplest form (such as 8/12, 10/18, 6/8, etc.) Create the cards by compiling a list of fifty fractions that the class should be able to reduce to lowest terms.
2. Blank BINGO game cards for each pupil (Cards may be reproduced on a copy machine and, if desired, may be laminated for repeated use.)
3. Pencils (or washable marking pens if using laminated cards) for each pupil
4. An ample quantity of beans or buttons
5. Three sets of "calling cards" with the letters **B I N G O** on them

B	I	N	G	O
1/2	3/4	2/3	1/8	5/6
1/4	3/8	1/6	2/3	5/8
1/3	1/8	3/4	7/8	1/2
5/6	1/3	7/8	5/8	1/4
5/8	1/5	3/5	1/2	1/6

Objectives:
As a result of this activity, the learner will:
- become more familiar with equivalent fractions.

Introduction:
1. Distribute cards and beans (or buttons) to the class.
2. In the blank BINGO-card boxes instruct the pupils to write the answers which they think will be called during the Equivalent BINGO game.
3. Emphasize the following points to the class: All the fractions the pupils choose to write on their BINGO cards must be reduced to lowest terms. Pupils are not allowed to write the same fraction more than once per column or row.
4. Explain to the class that you will call out a fraction that is *not* reduced to lowest terms. The object of the game is to attempt to find in the appropriate

Page 5.12   Activities for Elementary School Mathematics

column on their game cards the lowest-term equivalent of the fraction you call, and cover that fraction with a bean/button. Five across, down, or diagonally creates a BINGO.

Major Instructional Sequence:
1. Separately shuffle the BINGO calling cards and fraction cards, turn over the first card in each stack, and call out the combination resulting from both cards, for example, "B" 10/12.
2. If a pupil has a 5/6 in the "B" column, he or she can place a bean (or button) on the 5/6 *in that column only*. In any of the other columns, 5/6 may not be covered.
3. If the next number called is "I" 6/8, the pupil will look for 3/4 *in the "I" column only*. In any of the other columns, 3/4 cannot be covered.

B	I	N	G	O
1/2	3/4	2/3	1/8	5/6
1/4	3/8	1/6	2/3	5/8
1/3	1/8	3/4	7/8	1/2
5/6	1/3	7/8	5/8	1/4
5/8	1/5	3/5	1/2	1/6

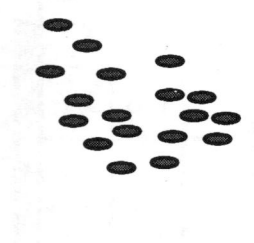

Given the example calls provided in the instructional sequence and the BINGO card illustrated on the previous page, a pupil would have the spaces above covered on his or her card so far.

4. As you call out each fraction, write its "lowest term" form in the appropriate column of a record sheet (to facilitate verification of BINGOs). See illustration below.

B	I	N	G	O
1/3	3/4		3/8	
2/3			1/3	

Closure or Evaluation:
The game continues until a pupil gets a BINGO and calls out, "Fraction BINGO!" Check to make sure answers are correct and in the proper column or row before pupils clear their cards. If there is in fact a BINGO, cards are cleared and the next round begins.

Activities for Elementary School Mathematics   Page 5.13

DECIMAL/FRACTION/PERCENT BINGO

Topic: EQUIVALENT DECIMALS, FRACTIONS, AND PERCENTS

Grade Level: 5-6   **Activity Time:** 1 class period

Materials Needed:
1. A list of commonly used decimals, fractions, and their equivalent percentages for each pupil (Decimals, fractions, and percents such as those listed below could be used.)

.50	1/2	50%	.33	1/3	33 1/3%
.75	3/4	75%	.25	1/4	25%
.875	7/8	87 1/2%	.125	1/8	12 1/2%
.20	1/5	20%	.67	2/3	66 2/3%
.625	5/8	62 1/2%	.375	3/8	37 1/2%
.10	1/10	10%	.167	1/6	16 2/3%

2. Individual "calling cards" with each of the decimals, fractions, and percents on the list
3. Blank BINGO game cards for each pupil (Cards may be reproduced on a copy machine and, if desired, may be laminated for repeated use.)
4. Pencils (or washable marking pens if using laminated cards) for each pupil
5. An ample supply of beans (or buttons)
6. Three sets of "calling cards" with the letters **B I N G O** on them

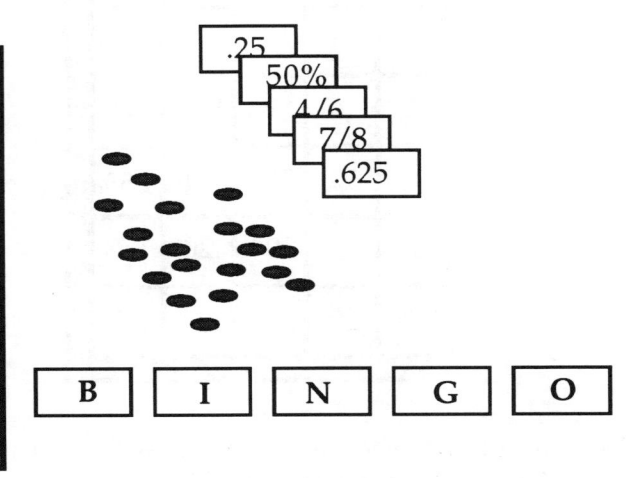

Page 5.14   Activities for Elementary School Mathematics

Objectives:
As a result of this activity, the learner will:
- become familiar with decimals, fractions, and percent equivalents.

Introduction:
1. Distribute BINGO playing cards and chips/beans/buttons (and marking pens, if the cards are laminated) to pupils.
2. Distribute the lists of commonly used decimals, fraction, and equivalent percentages to the class.
3. Instruct the pupils to write in the boxes on their cards the selections they think will be called during the BINGO game. No duplicate selections are allowed in the same row or column.

NOTE: *Tell the pupils that they are not permitted to have two possible correct responses in one column or row; e.g., I 50% and I .50 or B .25 and B 1/4.*

Major Instructional Sequence:
1. Separately shuffle the BINGO calling cards and fraction/decimal/percent calling cards, turn over the first card in each stack, and call out the combination resulting from both cards–for example, "B" 1/4.
2. If a pupil has a decimal or percent equivalent of 1/4 in the "B" column, he or she can place a bean (or button) on ONE correct equivalent *in that column only*. In any of the other columns, the equivalent may not be covered.
3. If the next number called is "I" 50%, the pupil will look for ONE fraction or percent equivalent of 50% *in the "I" column only*. In any of the other columns, the equivalent cannot be covered.

B	I	N	G	O
1/2	.33	2/3	.125	75%
25%	3/8	1/6	10%	5/8
1/3	.50	75%	7/8	25%
5/6	.167	50%	.20	1/8
5/8	1/5	.625	1/2	.375

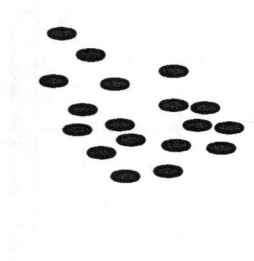

4. As you call out each equivalent, write it in the appropriate column on a record sheet (to facilitate verification of BINGOs). See illustration on the next page.

Activities for Elementary School Mathematics   Page 5.15

B	I	N	G	O
1/4	.20	3/4	75% .25	

Closure or Evaluation:
1. The game continues until a pupil gets five in a row, a column, or diagonally, and calls out "BINGO!"
2. Tell the pupils not to clear their cards until you verify that the answers are correct and in the proper column or row. It is also necessary to make sure the pupil did not cover any of the elements that were actually called out to the class rather than the elements' equivalents.

DECIMAL/FRACTION CONCENTRATION

Topic: FRACTIONS AND DECIMAL EQUIVALENTS

Grade Level: 5-6   **Activity Time:** 1 class period

Materials Needed:
1. Six sets of teacher-made decimal/fraction game cards (Samples of cards appear below. Each set should have 20 cards.)

| 1/2 | .50 | 1/4 | .25 | 7/8 | .875 |

NOTE: *Make sure each card you create has a "match;" e.g., 1/4 & .25, and the like.*

Objectives:
As a result of this activity, the learner will:
- be able to identify fractions and match them to their decimal equivalents.

Introduction:
1. Divide the class into six groups and distribute a set of cards to each group.
2. After discussing fractions and decimal equivalents, ask each group to place the game cards face down in rows of five across/four down.

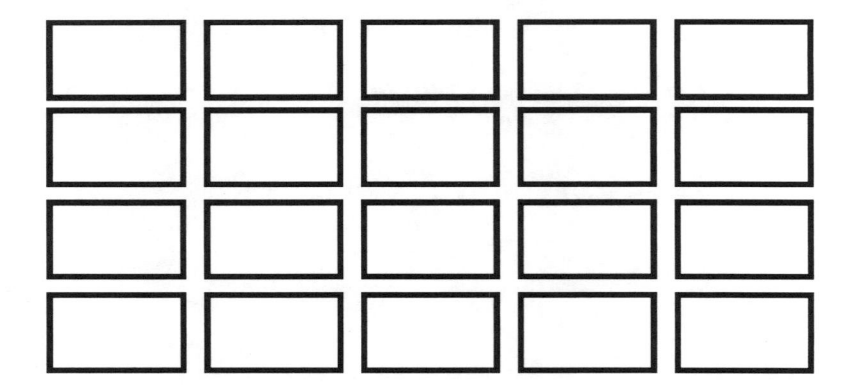

3. Explain to pupils that the object of the game is to match each fraction card with its equivalent decimal card.
4. Instruct pupils to try to remember the placement of the equivalent cards in order to be able to successfully match them at a later time. Make sure the class understands the game principles by drawing a sample "match" on the chalkboard. Then the first pupil in each group starts the game.

Activities for Elementary School Mathematics Page 5.17

Major Instructional Sequence:
1. The first pupil in each group turns over two cards, one at a time. If the cards do not match, they are returned to their original, face-down positions, and play continues to the right.
2. Pupils should be given enough time to see what has been turned up for future matches.
3. If a pupil matches two cards, such as the ones below, he or she gets to keep the cards.

$$\boxed{1/2} \quad \boxed{.50}$$

4. Circulate among the groups, monitoring play carefully, checking for understanding, and providing assistance when needed.

Closure or Evaluation:
The winner in each group is the pupil who has the most matched pairs at the finish of the game (when all of the cards have been matched).

Page 5.18   Activities for Elementary School Mathematics

DECIMAL TIC-TAC-TOE

Topic: CHANGING FRACTIONS TO DECIMALS

Grade Level: 5-6   **Activity Time:** 1 class period

Materials Needed:
1. 3 x 5 index cards with a common fraction on each of them
2. Blank tic-tac-toe cards, reproduced by a copy machine (one card for each group of four pupils)
3. Beans (or buttons) to cover answers on the game card

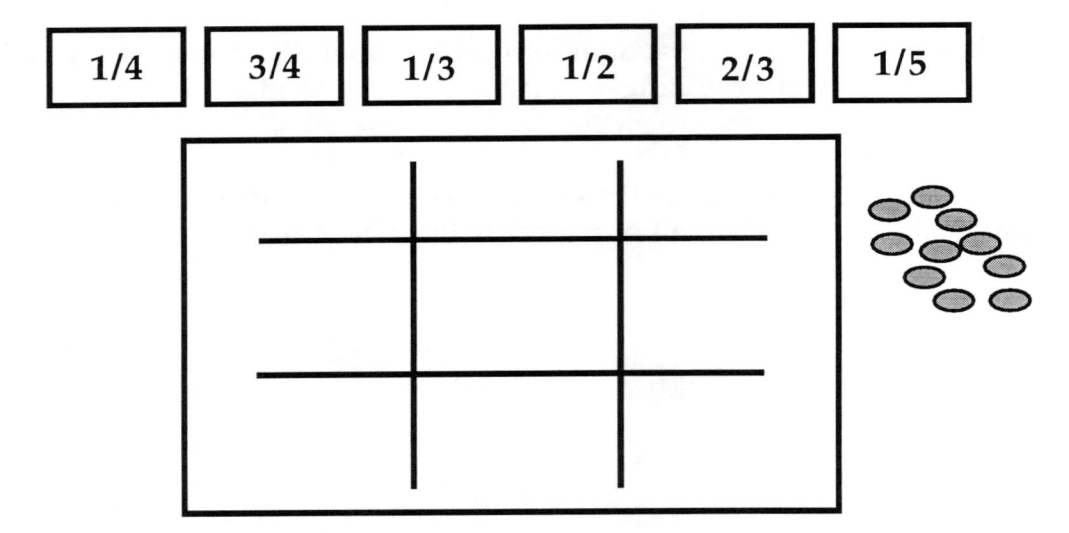

| 1/4 | 3/4 | 1/3 | 1/2 | 2/3 | 1/5 |

NOTE: *Cards could be laminated for repeated use. If you do so, provide erasable markers for each group to write decimal equivalents to the common fractions being studied.*

4. You may also provide to the class a list of fractions and their decimal equivalents (representative of the fractions on the index cards). Some pupils may not need the list; they will know the fractions and equivalent decimals from previous class discussions. A sample list follows.

1/2 = .5	1/3 = .333	2/3 = .667
3/4 = .75	1/5 = .2	3/5 = .6
1/6 = .167	1/8 = .125	3/8 = .375
1/20 = .05	5/8 = .625	1/9 = .111

Activities for Elementary School Mathematics   Page 5.19

Objectives:
As a result of this activity, the learner will:
- be able to match common fractions and decimal fractions.

In preparation for this activity, make a list on the chalkboard of the common fractions to be used in the game.

Introduction:
1. Call attention to the common fractions on the chalkboard. Let pupils volunteer to read each one aloud.
2. Divide pupils into groups of four.
3. One of the group of four players passes out the tic-tac-toe cards and the decimal equivalents lists to the other players. The pupils write the decimal equivalents on their cards from the list of common fractions on the chalkboard. Fifteen or twenty fraction cards are needed for the game.
4. Each group leader shuffles the fraction cards and places them in a stack.

Major Instructional Sequence:
1. First card is picked up by the leader and read to the group.
2. If a player has the decimal equivalent on his or her tic-tac-toe card, he/she marks it with a bean (or button). As an example, 1/3 is the first fraction called out to the group. On the card below, .333 would be marked if the pupil realized that 1/3 is equal to .333.

.333	.25	.20
.50	.667	.75
.125	.625	.875

3. The next card is turned up and the game continues with pupils covering the decimal equivalents on their cards.
4. Circulate among the groups, monitoring carefully, checking for understanding, and giving assistance when appropriate.

Closure or Evaluation:
The game continues until a player wins by getting 3 in a row, column, or diagonally.

CONCENTRATION TOSS

Topic: EQUIVALENT FRACTIONS

Grade Level: 5-6

Activity Time: 1 class period

Materials Needed:
1. Poster board divided into squares (three across and three down) with a simplest form fraction written on each square
2. Nine 3 x 5 index cards with an equivalent fraction on each card for the ones on the game board

1/3	3/4	2/5
5/8	1/6	1/8
2/3	1/4	1/2

6/9

   NOTE: *A second set of completely different equivalent fractions for the game board may also be necessary for later use.*

3. A small bean bag or chalkboard eraser

Objectives:
   As a result of this activity, the learner will:
   • become familiar with equivalent fractions.

Introduction:
1. Place the prepared poster board on the floor in front of the class and divide pupils into two teams.
2. Shuffle the nine cards and place them face down on a table in 3 rows of 3 per row. Before turning the cards face down, show them to pupils one at a time, reminding pupils to try and remember where each card is located.

Major Instructional Sequence:
1. If, for example, the bean bag/eraser lands on 1/2, the pupil must try to remember the position of the index card with the equivalent fraction on it for the 1/2 on which the bean bag landed. If the wrong card is turned up, it is returned to its face-down position and the next team takes a turn.
2. Teams take turns tossing the bean bag or eraser so it lands on the game board.
3. One point is scored for successfully matching two fractions.

Closure or Evaluation:
   The winning team is the one that makes the most matches per round.

CHAPTER SIX
GEOMETRY

This section incorporates plane and space figures. *Plane figures* can be drawn on a single, one-dimensional plane (for example, circle, square, rectangle, pentagon, triangle).

Space figures such as cube, prism, pyramid, and cylinder are perceived as three-dimensional and hollow.

Some textbooks refer to them as *geometric solids*. When introducing these concepts to pupils, it may help to use a square cut from cardboard and a cube, such as a block, to compare the two concepts, explaining that the square is flat (only one side is visible) but that the top and sides are visible on the cube.

Activities in this section cover area, perimeter, coordinate geometry, the geoboard, shape identification, and a geometric identification game that can be adapted to meet the needs of any class from first to sixth grade. This card game can be used to find the most precise name for a figure. A figure can be viewed in depth to find many of the geometrically correct names for the figure. Take the square, for example:

- The figure is a *closed curve* (you can start at one point and return to that point).
- It is a *simple closed curve* (you can start at one point and return without retracing any of the figure).
- A *polygon* is a simple closed figure made up of line segments.
- A *quadrilateral* is a simple closed figure made of four segments.
- The square is a special *rectangle* (it possesses all of the properties of a rectangle and is special because all four sides are the same length).
- A *parallelogram* is a quadrilateral with opposite sides parallel. (It could be considered a *rhombus* if the definition of rhombus contains no mention of the angles in the corners.)
- The *trapezoid* is also a possibility, depending on the definition. If the definition states that at least two sides are parallel, then the figure fits

this category. Some definitions state **only 2 sides parallel** (in which case the figure would not fit this definition of trapezoid).

Remember that *square* is the most precise name. It eliminates many of the other figures that could be included under other terms.

Another activity in this chapter, Triangle Sort, explores the triangle in great detail. It classifies the triangle by its angles.

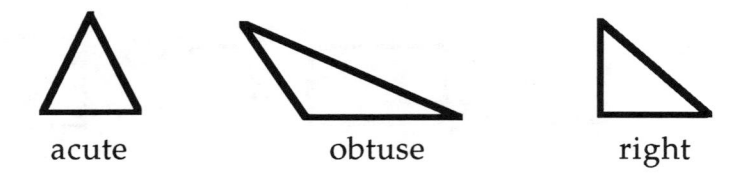

acute   obtuse   right

The game also classifies triangles by the length of the sides.

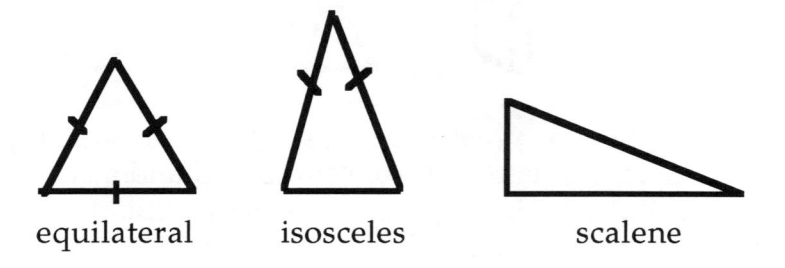

equilateral   isosceles   scalene

If all three sides are congruent, the triangle is an equilateral triangle. If only two sides are the same length, it is an isosceles triangle. (Some books say at least two sides must be congruent; therefore the equilateral triangle would be included in the isosceles category.) Scalene triangles have no sides the same length.

Activities for Elementary School Mathematics   Page 6.3

GEO-BOARD SHAPES

Topic: GEOMETRY

Grade Level: 1-6   **Activity Time:** 1 class period

Materials Needed:
   1. A geo-board for each group of two to three pupils
   2. Rubber bands
   3. Paper and pencil for each pupil

Objectives:
   As a result of this activity, the learner will:
   - become familiar with different plane geometry shapes.
   - be able to find area and/or perimeter of plane figures.

Introduction:
   1. Use the geo-board for simple shape recognition activities with younger pupils. Simple figures like square, rectangle, and triangle can be illustrated on the geo-boards for the early grades.

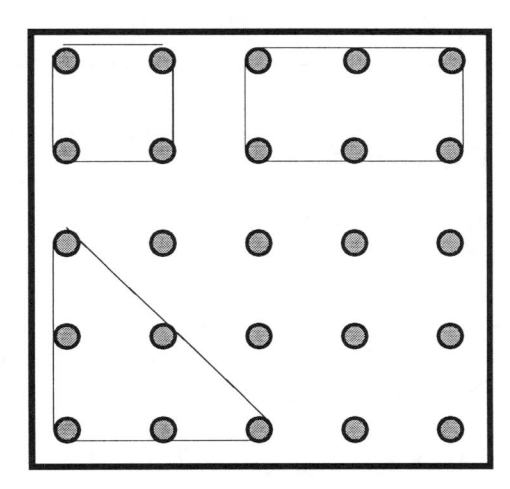

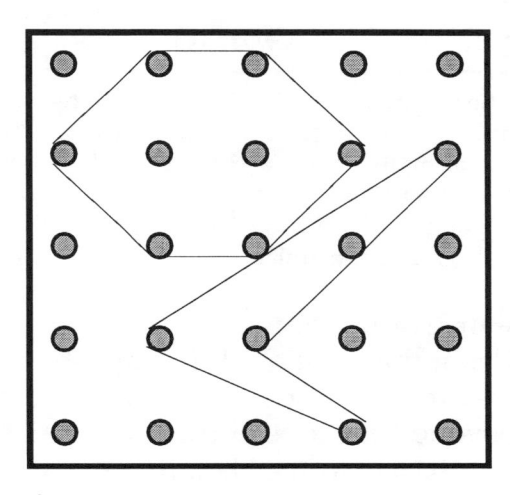

   2. Other figures, such as the hexagon and concave quadrilateral, can be used in the upper elementary grades, for whom area and perimeter can also be demonstrated.

Major Instructional Sequence:
   1. Before class, create several geometric figures on the geo-board by placing pegs in the pegboard and stretching a rubber band around them to create the outline of the figures.

2. When teaching a lesson on area and perimeter on the geo-board, find the area of squares and rectangles first. Explain to the class that on the geo-board, the area of a figure equals length times width counting the *spaces*, not the pegs, *along the sides of the figure.*
3. After the class understands the method for finding the area of a square, explain the way to figure out the area of a triangle as 1/2 ab. This is actually half of the length of the base of the triangle (*a*) times its height (*b*). It is half the area of a rectangular region with the same measurements. Therefore, half of the rectangle's area is the triangle's area. This is true for all triangles.

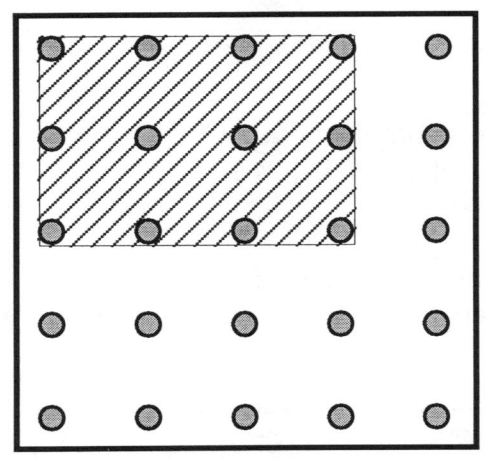

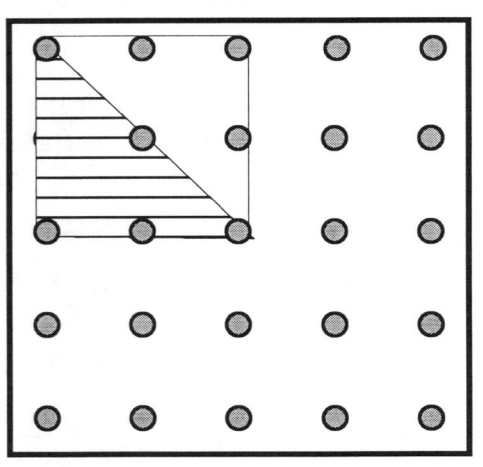

The rectangle above is 3 spaces long by two spaces wide, 3 x 2 = 6 square units. The area of a square is found the same way as the area of the rectangle. A triangle's area is found by the formulas 1/2 ab, so the length of its base is found and multiplied by its height and then divided by two so half of the area is found. The triangle is 1/2 of the rectangle, which is 2 x 2 or 4 square units, so the area of the triangle is 1/2 of 4 square units or 2 square units.

Closure or Evaluation:
1. Using the geo-board, prompt the class to find the perimeter of a triangle. (Perimeter can be found by using the Pythagorean theorem, but this may be beyond the scope of the sixth grade.)
2. Emphasize to the class that the area of *ANY* triangle can be found by finding 1/2 *ab*! When looking for the area of the following figures, finding the triangle's area first will be helpful.

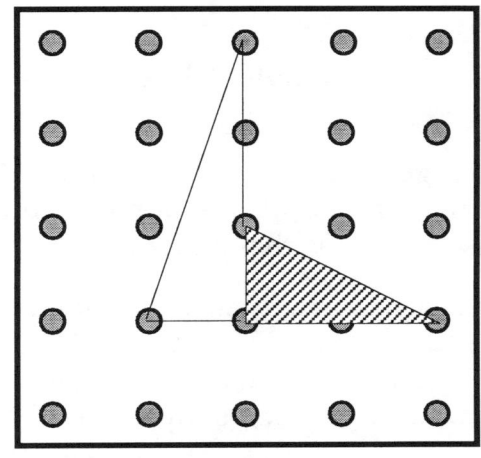

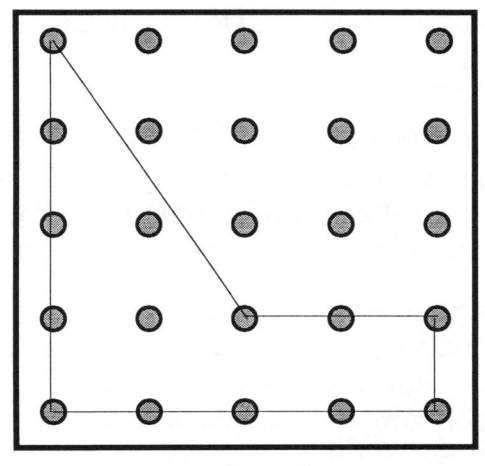

Activities for Elementary School Mathematics — Page 6.5

figure 1   figure 2

The area of figure 1 above can be found by finding the area of two triangles. One triangle is 1 x 3, so its area is 1 1/2. The other triangle is 1 x 2, so its area is 1, making a total area of 2 1/2 square units. Figure 2's area can be found by finding the area of the rectangle on the bottom of the figure, which is 4 x 1, and the area of the triangle on the top of the figure can be found by taking 2 x 3 and then dividing by 2. Add 4 and 3 to get 7 square units for the area of figure 2.

Page 6.6   Activities for Elementary School Mathematics

GEO-CARD GAME

Topic: BASIC GEOMETRIC CONCEPTS

Grade Level: 3-6   **Activity Time:** 1 class period

Materials Needed:
1. Two sets of geometric cards (3 x 5) for each group of four to five pupils (The first set contains geometric terms with which the pupils are familiar and could be varied to cover areas to be stressed, such as the following:)

squares	four-sided figures
quadrilaterals	all 90° corners
triangles	three-sided figures
polygons	simple closed figures
closed figures	trapezoids
rhombuses	octagons
pentagon	hexagon
open figures	not closed figures
parallelogram	ellipse (often referred to as an oval)
circles	all points equidistant from center

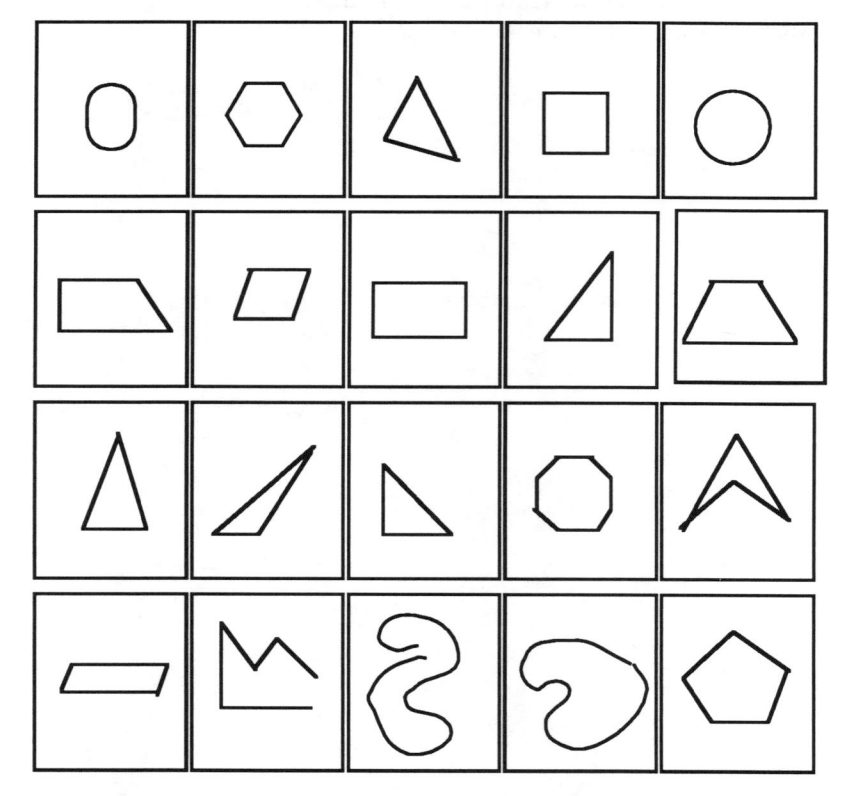

The second, corresponding set of cards for the sample list could look like this.

Activities for Elementary School Mathematics   Page 6.7

Objectives:
As a result of this activity, the learner will be able to identify:
- appropriate geometric shapes.
- geometric shapes with several correct geometric terms.
- the most precise term for a figure.

Introduction:
1. Divide pupils into groups of four to five.
2. Place the geometric terms cards face down on the table in one stack.
3. Place all of the geometric figure cards face up on the table.

Major Instructional Sequence:
1. The first player in each group turns up the first terms card and tries to identify all the figure cards that match the term card. If a player turned up "square," for example, he or she would correctly select:

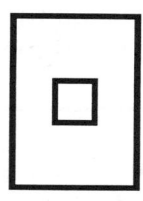

2. The group discusses the selection to make sure all agree on the answer.
3. Circulate among the pupils, checking for understanding and giving assistance as necessary.

Closure or Evaluation:
Play continues until all term cards have been turned up and discussed.

NOTE: *Be aware of differences in some of the terms as defined by various books. For example, some books define a rhombus as follows:*

a) a quadrilateral with four congruent sides and four oblique (neither parallel nor perpendicular) angles.
b) a quadrilateral with four congruent sides.

Answers for a) above   Answer for b) above

Some other differences are shown below.

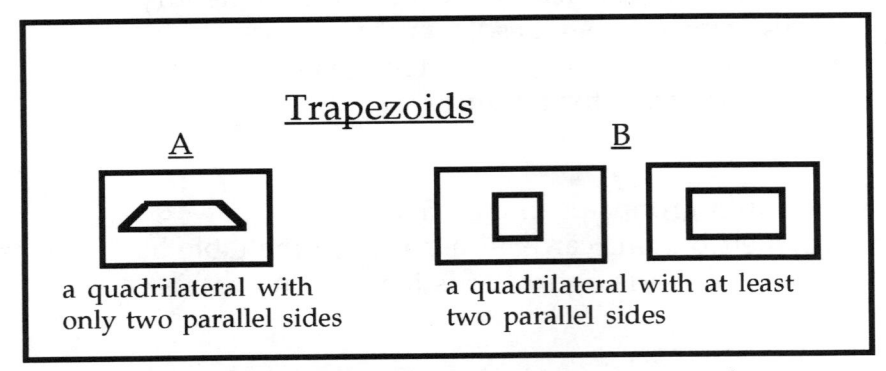

```
                        Trapezoids
            A                            B
         [figure]                  [figure]  [figure]
    a quadrilateral with      a quadrilateral with at least
    only two parallel sides   two parallel sides
```

In the definition for B above, squares and rectangles would be included but would not be part of A.

Some quadrilaterals, like the one below, may cause confusion among pupils because of their odd shapes. Remind pupils that a quadrilateral is a closed figure (called a polygon) with four sides, and the figure below fits that description:

Triangles can present a problem in that isosceles triangles can be defined as either a triangle with two sides congruent or a triangle with *at least* two sides congruent:

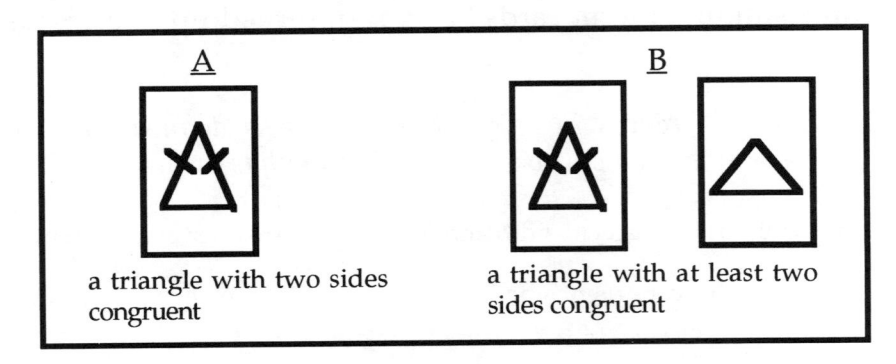

```
            A                            B
         [figure]                  [figure]  [figure]
    a triangle with two sides   a triangle with at least two
    congruent                   sides congruent
```

Be aware of how the books in your classroom define these terms in order to help pupils avoid misunderstandings.

Activities for Elementary School Mathematics   Page 6.9

SHAPE-O

Topic: GEOMETRIC SHAPES AND COLORS

Grade Level: 1-6   **Activity Time:** 1 class period

Materials Needed:
  1. Forty 3 x 5 index cards with geometric shapes of different colors on each

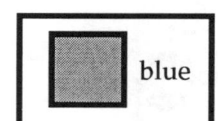

     blue   yellow   red   green

     (The cards should have on them a minimum of 6 different geometric shapes of at least 6 colors.)
  2. Pupil game cards with 4 columns of 4 squares, each column with a different color listed above it and four different shapes down the side, to the left of each row (Each game card should be different.)
  3. Beans (or buttons) for markers

Objectives:
  As a result of this activity, the learner will be able to:
  • use a grid system.
  • recognize different geometric shapes.
  • recognize different colors.

Introduction:
  Explain to the class that you will be calling out different-colored geometric shapes. Instruct the pupils to check their game cards to see if they have each shape on their cards in the column of the color you called. They will use the beans as markers to place on their game cards when they have matches.

Major Instructional Sequence:
1. Mix the index cards and stack them face down on your desk.
2. Turn up the first card and call it out to class (for example, "red square").
3. Continue play by calling out other shapes of different colors, such as a green hexagon.

The player using the card below would not be able to mark a green hexagon. If a green triangle were called, however, it could be marked.

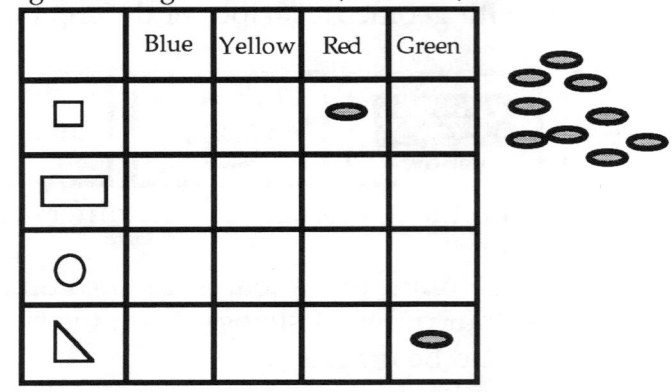

	Blue	Yellow	Red	Green
□			●	
▭				
○				
△				●

Closure or Evaluation:
1. The game ends when a pupil who has four across, down, or diagonally calls out, "SHAPE-O!"
2. The winner of any game may become the caller for the next game, if desired.

Variation:
The game can be varied to fit the ability of the pupils involved in the activity. For example, a sixth-grade geometry lesson could utilize game cards that look like this:

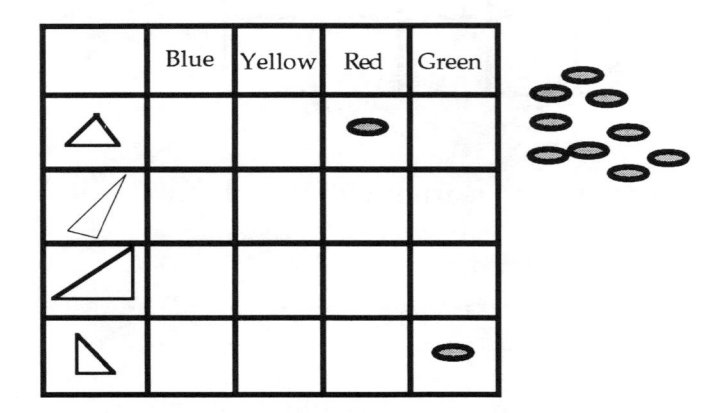

	Blue	Yellow	Red	Green
△			●	
◁				
◣				
◿				●

In this case, the index cards you call out to the class would include such geometric shapes as a scalene triangle, an acute triangle, an isosceles triangle, and an equilateral triangle.

Activities for Elementary School Mathematics

WHAT SHAPE?

Topic: GEOMETRIC SHAPES

Grade Level: 4-6　　　　　　　　　　　　　　　**Activity Time:** 1 class period

Materials Needed:
1. Containers from around the home and school (such as an oatmeal box, cracker box, pizza box, large tea-bag box, and other unusual items made of cardboard, paper, and the like, that can be broken down easily and safely)
2. Paper and pencil for each pupil

Objectives:
As a result of this activity, the learner will:
- have a better understanding of space figures (three-dimensional, hollow figures) occurring in everyday life.

Introduction:
Hold up an item for the class to observe, such as a cylindrical oatmeal box or a milk carton. Explain to pupils that they will attempt to sketch what they think the "opened-up" (one-dimensional, or flattened out) container will look like. (See the illustrations below.)

Major Instructional Sequence:
1. Continue to hold up the object in front of the class. Turn it in as many directions as is necessary to display its top, sides, etc.
2. Ask the pupils, "If this box (or carton, etc.) was opened up and spread out flat, can you imagine what it would look like?"
3. Instruct pupils to sketch their ideas. Keep the items displayed so that any pupil who feels he or she needs to look at an item again may do so.

Page 6.12   Activities for Elementary School Mathematics

Closure or Evaluation:
1. After the class has made sketches of the displayed items, the items can be opened up, flattened, and discussed as to variations between pupils' sketches and the actual, "opened-up" arrangements.

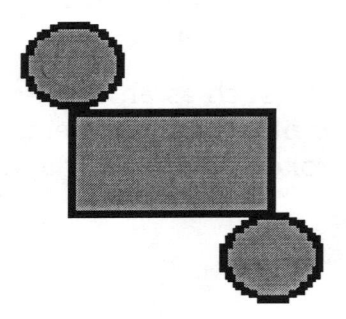

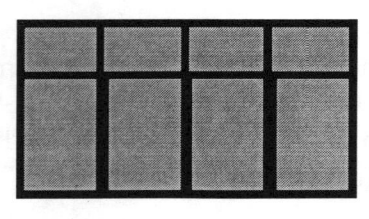

Activities for Elementary School Mathematics

PERIMETER

Topic: PERIMETER

Grade Level: 4-6

Activity Time: 1 class period

Materials Needed:
1. A 3 x 5 inch index card for each group of 3 or 4 pupils
2. An 8 x 11 sheet of heavy construction paper for each group
3. A three-yard-long section of string for each group (long enough to go completely around the 8 x 11 paper and the index card, with some excess)
4. A ruler or yardstick for each group
5. Scissors for each group

Objectives:
As a result of this activity, the learner will:
- become familiar with the concept of perimeter.

Introduction:
1. Divide pupils into groups of 3 or 4.
2. Distribute all of the materials to each group.
3. Explain that the first activity will be taking the string and placing it around the borders of the index card as closely as possible. Pupils from each group should place string as in the illustration below.

4. Instruct the pupils to mark the string where it meets, cut the string, and then measure it. (The length will be 16 inches for a 3 x 5 index card.)

NOTE: *Do not use the word perimeter until concept is understood as the distance around a polygon.*

Major Instructional Sequence:
1. Repeat the procedure for the 8 x 11 sheet of paper.
2. Discuss the procedure with the class, emphasizing that the lengths of string

have measured the distance around the edges of the paper and of the card. It may be helpful to use words such as *border* or *boundary* in the discussion.
3. When the class has a basic understanding, ask the pupils whether they can determine a way to find the distance around the borders of a rectangle without actually measuring the entire way around the rectangle. (Measure halfway and multiply by two.) Explain that the formula for finding this distance around a rectangle is *P = 2 (l + w)*.

Closure or Evaluation:
1. At this point tell the class that they have found the *perimeter*, the distance around polygons (simple closed figures made of line segments).
2. Lead the class in discussion about finding the perimeter of a square *(P = 4s)*.

Activities for Elementary School Mathematics

AREA

Topic: AREA

Grade Level: 4-6 Activity Time: 1 class period

Materials Needed:
1. A 3 x 5 index card for each group of 3 or 4 pupils
2. A 5 x 7 index card for each group
3. 20-25 one-inch-square pieces of construction paper for each group

Objectives:
As a result of this activity, the learner will:
- become familiar with the concept of area.

Introduction:
1. Divide pupils into groups of 3 or 4.
2. Distribute all of the materials to each group.
3. Tell the class to begin the activity by placing the squares on the 3 x 5 card. Explain that the pupils should place the squares next to each other, not overlapping them, so that the top of the index card is completely covered. After it is covered, ask the pupils to count the squares.

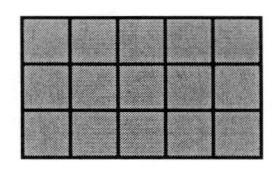

NOTE: *Do not use the word area until concept is understood as the surface of the polygon.*

Major Instructional Sequence:
1. Ask the groups to repeat the procedure for the 5 x 7 index card.
2. When the pupils discover that they do not have enough squares to cover the entire 5 x 7 rectangle, ask them if they can discover a way to find "the amount of space inside the rectangle" or how much space it "takes up."
3. Circulate among the pupils, checking for understanding, and giving assistance as necessary.

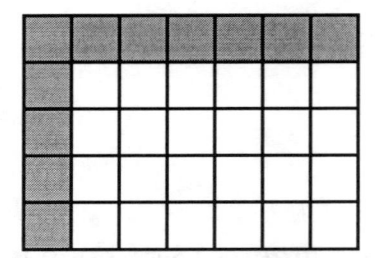

Page 6.16   Activities for Elementary School Mathematics

Place seven squares along one side and five along the other. This indicates we could place five rows of seven squares on top of this card, if we had enough squares.

Closure or Evaluation:
At this point, tell the class they have discovered the *area*, the surface of a polygon. The formula for determining the area of a square is $A = l \times w$.

Activities for Elementary School Mathematics

Page 6.17

EVERYDAY GEOMETRY

Topic: GEOMETRY

Grade Level: 4-6

Activity Time: 1 class period

Materials Needed:
1. Cereal boxes, oatmeal boxes, and other food containers of various shapes (made of material such as cardboard that pupils will be able to cut with scissors)--one container for each pupil
2. Two large sheets of paper and a pencil for each pupil
3. Scissors for each pupil

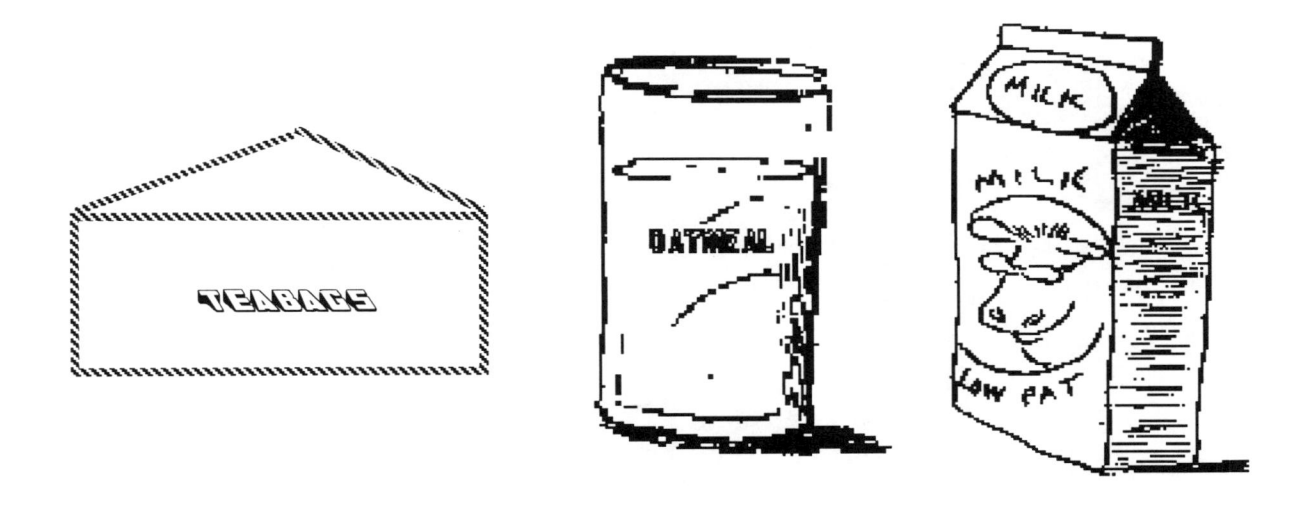

Objectives:
As a result of this activity, the learner will be able to:
- count the edges, faces and corners of space figures.
- identify different regional shapes which make up each face.

Introduction:
1. Distribute the containers to the class.
2. Ask each child to count the number of edges, faces, and corners on his or her container and to record this information at the top of a sheet of paper.
3. Instruct the pupils to trade containers and verify each other's findings.
4. Each individual finding should be recorded at the top of a separate sheet of paper.

Major Instructional Sequence:
1. Instruct pupils to cut along the edges of their containers in such a way as to keep the container in one piece. Circulate among the pupils, checking for

understanding, and providing assistance as needed.
2. The container should be flattened and then traced around on the sheet of paper on which the pupils recorded the findings for that container. Ask pupils to make the same trade of containers once again so that they each have two tracings with corresponding information.
3. Instruct pupils to separate and color each face on their tracings a different color and identify each plane region.

Closure or Evaluation:
1. Show other containers to the class and ask them to predict how many faces, edges and corners each will have.
2. Shapes of the faces could be guessed as you hold each container up in front of the class, then verified by group analysis and examination.

Activities for Elementary School Mathematics

GEO-SORT

Topic: GEOMETRIC SHAPE IDENTIFICATION

Grade Level: 5-6

Activity Time: 1 class period

Materials Needed:
1. Geometric regions like the ones illustrated below, to be cut out of cardboard or poster board:

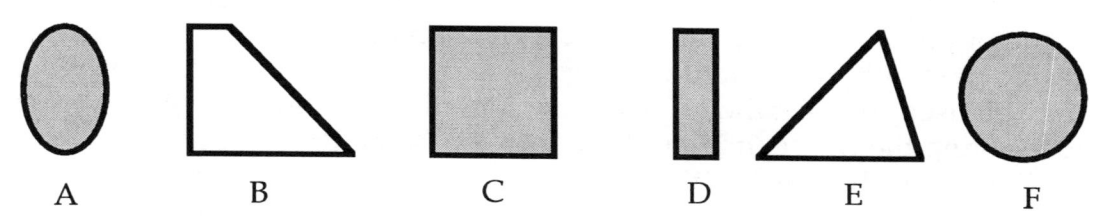

   A   B   C   D   E   F

   a. oval (some books use ellipse as the best name)
   b. closed curve, simple closed curve, polygon, quadrilateral or trapezoid
   c. closed curve, simple closed curve, polygon, quadrilateral, rhombus, rectangle or square
   d. closed curve, simple closed curve, polygon, quadrilateral or rectangle
   e. closed curve, simple closed curve, polygon or triangle
   f. closed curve, simple closed curve or circle

   NOTE: definitions vary from text to text, so make sure you use the definitions which are used in the class text.

2. A Game Bag (any opaque bag large enough for a pupil to put a hand into)

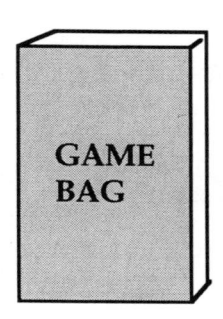

   GAME BAG

3. A ruler and protractor

Objectives:
As a result of this activity, the learner will:
- be able to classify geometric shapes in several different ways.
- realize that geometric shapes can have several names.

Introduction:
1. Before class begins, prepare cutout figures and place them in the Game Bag.

2. Emphasize that figures are made up of line segments, except for the circle and the oval. Tell the pupils that they will be considering the outer edges of the regions only.
3. Tell the class that they will take turns closing their eyes and selecting a figure from the Game Bag. (Caution pupils not to look at their figures.)

Major Instructional Sequence:
1. A pupil reaches into the Game Bag and selects a figure. Without looking at it, the pupil tries to identify the type of figure selected. It may be identified by any of the names that are geometrically correct.
2. If the pupil needs to look at the figure, let him or her do so only after trying to identify it without looking. If necessary, a ruler and/or protractor may be used to assist the pupils in the identification process. After identifying the figure, it is put back into the Game Bag.
3. Walk around the class, letting each pupil take his or her turn.

Closure or Evaluation:
1. The game continues until all pupils have attempted to identify at least one figure from the Game Bag.
2. If teams are used, one point could be scored for each correct identification.

Activities for Elementary School Mathematics   Page 6.21

TRIANGLE SORT

Topic: TRIANGLE IDENTIFICATION

Grade Level: 5-6   **Activity Time:** 1 class period

Materials Needed:
1. Triangular regions like the ones illustrated below, cut out of cardboard or poster board

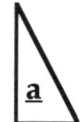

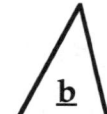

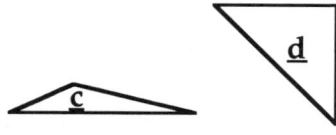

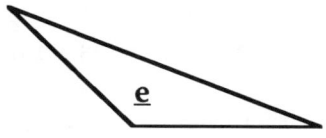

   a. right scalene triangle
   b. acute scalene triangle
   c. obtuse scalene triangle
   d. right isosceles triangle
   e. obtuse isosceles triangle

   Note: The appendix includes reproducible triangles of each type.

2. A Game Bag (any opaque bag large enough for a pupil to put a hand into)

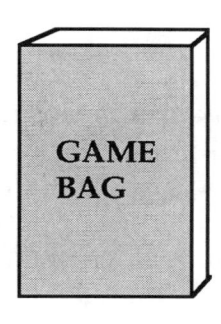

3. A ruler and protractor

Objectives:
   As a result of this activity, the learner will:
   - classify triangles by the degree of their angles.
   - classify triangles by the lengths of their sides.
   - recognize that triangles can be classified both by degrees of their angles and by lengths of their sides.

Introduction:
1. Before class, cut out the triangles and put them into the Game Bag.
2. Identification of the triangles used for the game preferably should come from the definitions in your textbook in order to avoid confusion among the pupils.

Sample classifications are as follows:
Classification by angles:
- Acute triangles have all angles less than 90 degrees.
- Obtuse triangles have one angle greater than 90 degrees.
- Right triangles have one angle of exactly 90 degrees.

Classification by sides:
- Equilateral triangles have all three sides congruent.
- Scalene triangles have no sides congruent, all three sides are of different lengths.
- Isosceles triangles have two sides which are congruent.*

*NOTE: *Definitions vary from text to text. Some texts use the definition "at least two sides congruent," while other texts define isosceles triangles as having "only two sides congruent." Decide in advance which definition to use for the game (preferably the one that pupils have been using in their texts and in class). Should you choose the definition which states "at least two sides congruent," also include equilateral triangles in this group.*

3. Emphasize to the class that triangles are made up of line segments which outline their regions and that pupils will be considering the outer edges of the triangular regions only.

Major Instructional Sequence:
1. Tell the class that they will take turns closing their eyes and selecting a figure from the Game Bag. Caution pupils not to look at their figures.
2. A pupil reaches into the Game Bag and selects a triangle. Without looking at it, the pupil tries to identify the type of triangle selected. It may be identified by either the length of sides or the degrees in the angles, or both.
3. If the pupil needs to look at the triangle, let him or her do so only after trying to identify it without looking. If necessary, a ruler and/or protractor may be used to assist the pupils in the identification process. After identifying the triangle, it is put back into the Game Bag.

Closure or Evaluation:
1. The game continues until all pupils have attempted to identify at least one triangle from the Game Bag.
2. If teams are used, one point could be scored for each correct identification of the sides or the angles, with two points possible for correctly identifying each triangle. A bonus point can be awarded if the pupil can identify the triangle by both attributes without looking at it first.

Activities for Elementary School Mathematics

BATTLE SHIP

Topic: COORDINATE GEOMETRY

Grade Level: 5-6

Activity Time: 1 class period

Materials Needed:
1. Two game cards per player, created to resemble the following illustration:

	a	b	c	d	e	f	g	h	i	j	k
1											
2											
3											
4											
5											
6											
7											
8											
9											

Note: *The appendix includes a reproducible game card like the ones above.*

2. Red and blue colored pencils for each player
3. A regular pencil for each player

Objectives:
As a result of this activity, the learner will:
- become proficient in coordinate geometry.

Introduction:
1. Separate the class into groups of two. Pass out two game cards per player (four cards per pair of pupils).
2. On one of their two cards, instruct the pupils to use their regular pencils to draw rectangles which represent ships (their "boat cards"). The size of the ships (i.e., how many squares may be covered per ship) and how many ships may be drawn per card will be determined by the amount of time available to play the game.
3. Caution pupils to keep the drawings on their "boat cards" hidden from their opponents.

Page 6.24 Activities for Elementary School Mathematics

Major Instructional Sequence:
1. On the second of the two cards, explain to the pupils how they will record the shots they take at their opponents' ships.

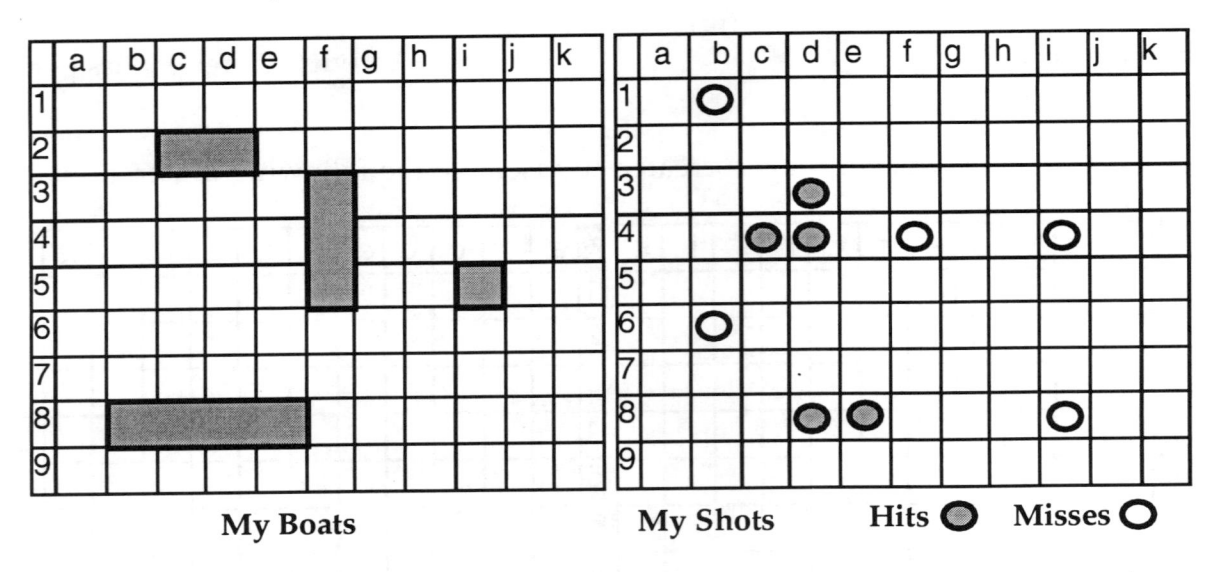

	a	b	c	d	e	f	g	h	i	j	k
1											
2			■	■	■						
3						■					
4						■					
5						■			■		
6											
7											
8			■	■	■	■	■				
9											

My Boats

	a	b	c	d	e	f	g	h	i	j	k
1		○									
2											
3				●							
4			●	●		○			○		
5											
6		○									
7											
8				●	●				○		
9											

My Shots   Hits ●   Misses ○

On the first card, boats are represented by dark, rectangular shapes. On the second card, the clear ovals represent misses (to be marked in blue) and the dark ovals represent hits (to be marked in red). Pupils in each pair take turns calling out coordinates. If a hit is made, the opponent must say so immediately so that the caller can mark his/her card with a red oval. (It is a good idea for the opponent receiving the hit to also mark his/her boat card to keep track of the condition of his/her boats.) If no hit is made, the shot is marked with a blue oval.

Closure or Evaluation:
1. When all the squares in the grid that represent a player's ship have been completely filled in by an opponent's hits, the player says, for example, "You sank my two boat" (i.e., the boat that was located predominantly in row two) or "You sank my F boat" (located predominantly in column F).
2. The game goes on until all of one partner's ships have been sunk or, if time is short, whoever has the most hits during a set time period wins.
3. Explain that for this game the coordinates are the *spaces* on a grid that make up a game card, and in actual coordinate geometry the coordinates are the intersections of *lines*.

Activities for Elementary School Mathematics   Page 6.25

PI, ANYONE?

Topic: GEOMETRY, *PI*

Grade Level: 6   **Activity Time:** 1 class period

Materials Needed:
1. The top of a coffee can and three other circular objects of different sizes, such as a coke can and a round cup or glass
2. Masking tape
3. Chalkboard

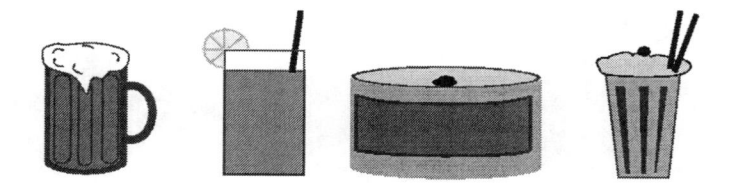

Objectives:
As a result of this activity, the learner will:
- better understand of the value of *pi* (approximately 3.14)

Introduction:
1. Put masking tape around the circular edge of each object that you have selected. (On a glass, cup, or any other such three-dimensional object, emphasize the fact that you are looking at *the top of the object only*.) Allow pupils to inspect your work.

Major Instructional Sequence:
1. Show the objects to the class and tell them that you have taped the circumference of the circles.
2. Remove the tape strips from each object and place them horizontally on the chalkboard.
3. Place an object on its corresponding tape, beginning at the left-hand side. Then show the class how the circular part of it fits three times and a little more across the tape. Repeat the demonstration with the other objects.

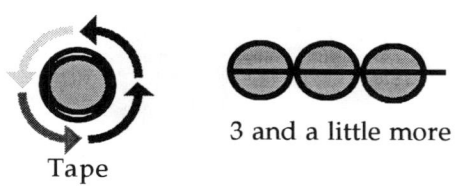

Tape   3 and a little more

4. Emphasize that each object fits three times and a little more across the tape each time you measure.

Closure or Evaluation:
1. Tell the class that they have just discovered *pi*.
2. Explain that the circumference of any round object is three and a little more diameters.

> **NOTE:** *Pi* is approximately 3.1415927. Some books use 3.1, others 3.14, and still others 22/7. All are approximate, since *pi* does not terminate or repeat.

CHAPTER SEVEN
ESTIMATION

Estimation skills should be built into the curriculum with whole numbers, operations, and measurements so that children understand estimation by the time they reach third grade. Children should be encouraged to guess and then check their guesses. Pupils should evaluate their estimates, and then they should adjust their conjectures in light of these checks or evaluations.

One of the activities deals with the amount of water that can be put on a penny (or a nickel, dime, or quarter). The reverse side of the coin could also be a part of the experiment in estimation to see what difference the configuration of the coin's surface would make on the amount of water a face would hold. The size of the coin, whether it is new or old, and whether it is heads or tails should be considered or experimented with in the classroom. What happens when a small amount of soap is placed on a coin face could also be considered. (Soap breaks the surface tension of water causing it to immediately drain off the coin.)

The value of money and the amount of purchase is another situation in which pupils may practice estimation. Pupils should learn that a good shopper needs to be able to estimate the cost of items that are going to be purchased if a limited budget is involved. Following is an easily understood example:

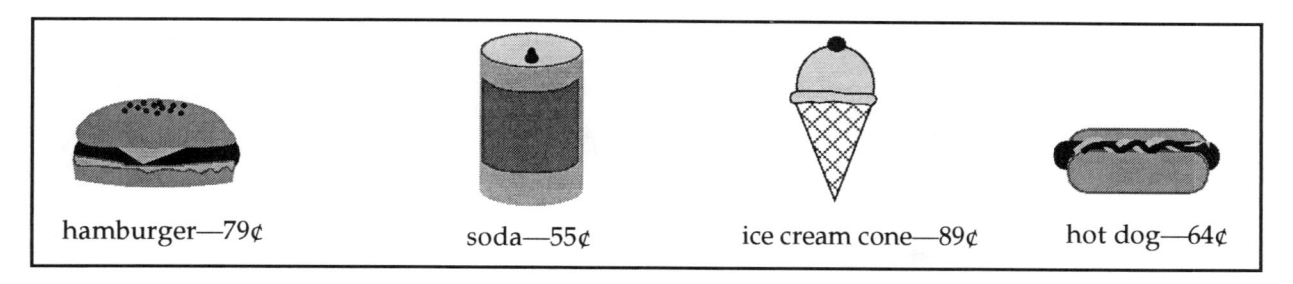

hamburger—79¢   soda—55¢   ice cream cone—89¢   hot dog—64¢

Does a pupil with $3 for lunch have enough to buy a hot dog, two hamburgers and a soda?
Estimate and then find the actual cost of lunch.

Estimating lengths of objects is also addressed. The pupils must find items they estimate to be a certain length and compare their estimates to actual measurement. Rulers, yardsticks, or meter sticks are good models to start with in the classroom when estimating lengths. Then use a variety of objects as models, and instruct pupils to find items that are the same length as the model item. Use whatever objects are handy and can serve as a physical model for the pupils.

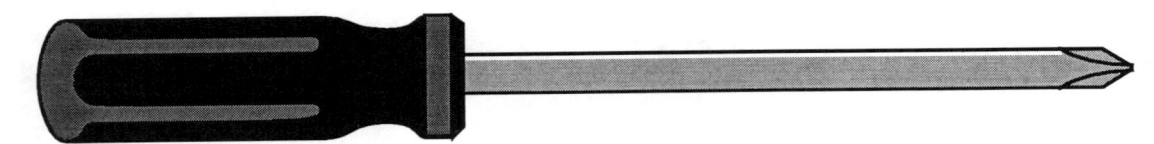

The screwdriver is approximately six inches long. What can you find around the classroom that is about the same length as the screwdriver? Can you find something in the room that is twice as long as the screwdriver?

The activities in this section contain a wide variety of topics addressing a pupil's ability to estimate in different situations. Lengths, costs, amount of items in a container, number of blinks of the eye during a certain time period, and other activities are presented to help pupils improve their estimation skills.

Activities for Elementary School Mathematics

Page 7.3

OUTFIT ESTIMATE

Topic: ESTIMATION AND MULTIPLICATION

Grade Level: 2-3

Activity Time: 1 class period

Materials Needed:
1. A cutout set of four T-shirts and four pairs of shorts (each differently patterned) for each group of four pupils

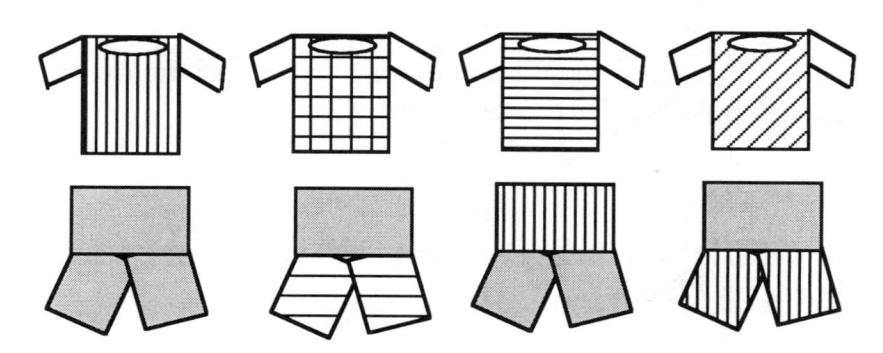

2. Paper and pencil for each pupil
3. Crayons for each group

Objectives:
As a result of this activity, the learner will:
- estimate how many different jogging outfits could be put together
- multiply to find out the exact number of jogging outfits.

Introduction:
1. Divide pupils into groups of four.
2. Distribute sets of T-shirts and shorts to each group of pupils.
3. Distribute the paper, pencils and crayons.

Major Instructional Sequence:
1. First instruct the pupils to *estimate* and record how many different outfits they think a jogger could make if the jogger had four T-shirts and four pairs of shorts.
2. Each group should put together shorts and T-shirts to make outfits and then copy the outfits and color them on their papers.

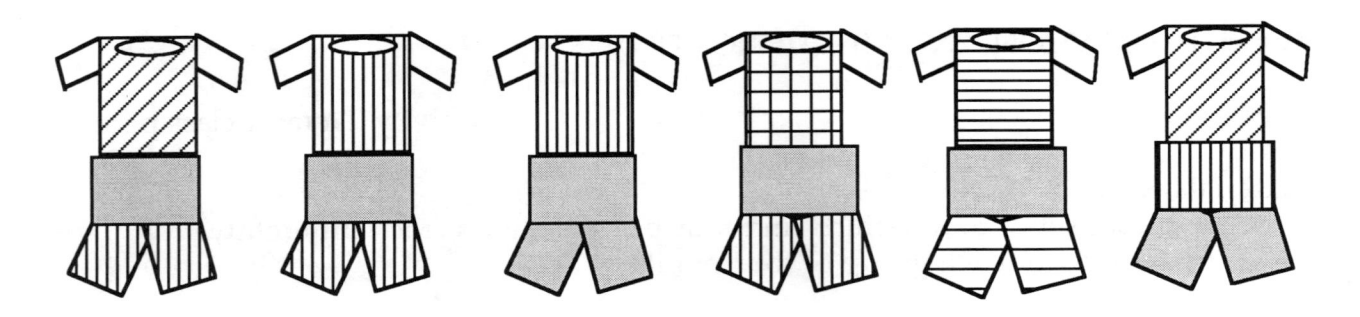

Page 7.4   Activities for Elementary School Mathematics

3. The class should discuss their outcomes and arrive at the exact number of outfits that was possible for the problem.

Closure or Evaluation:
Other combinations would be given out next, such as six T-shirts and three pairs of shorts, and the like.

Activities for Elementary School Mathematics   Page 7.5

CLASS ESTIMATE

Topic: ESTIMATION

Grade Level: 2-4   **Activity Time:** 1 class period

Materials Needed:
1. A large bag of any bite-size snacks, such as cheese puffs, M & M's, or jelly beans
2. A 3 x 5 card for each pupil in the class

NOTE: *When choosing edible items for estimation, you may want to take the neatness factor into account (e.g., M&Ms, jelly beans, and cheese puffs are less messy than some other choices).*

Objectives:
As a result of this activity, the learner will:
- improve ability to estimate.

Introduction:
1. Show the class the bag of treats (in this case, cheese puffs).
2. Distribute the cards to each pupil.

Major Instructional Sequence:
1. Let the class see the size of the bag and the size of the cheese puffs, but do not let them discuss their estimates among themselves.
2. Ask the pupils to put their names on their cards and number down the card to four (or whatever amount represents the number of weeks you will be working on estimation). Instruct the class to record on their cards their estimates of the number of cheese puffs in the bag.
3. Collect the cards and pin them on the bulletin board.

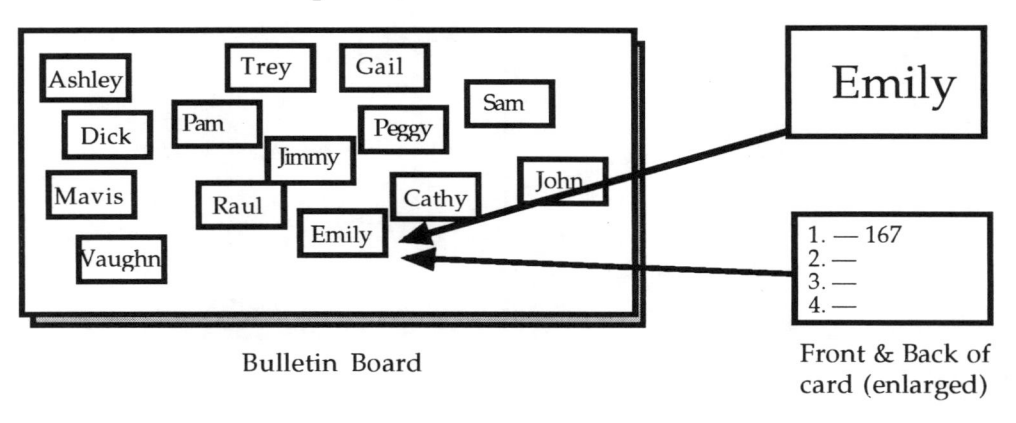

Bulletin Board   Front & Back of card (enlarged)

Closure or Evaluation:
1. On Friday have several pupils carefully count the cheese puffs. The pupil with the closest estimate without going over the actual amount is the winner. Then distribute the cheese puffs and enjoy a little Friday treat!
2. Use the M & M's the following week and follow the procedure outlined above, but do not let the class know in advance what the estimation activities are going to be.
3. In five or six weeks, try another, smaller or larger bag of cheese puffs to see if the class estimates are closer to the actual number this time.
4. Try objects in different amounts and shapes each week, for example, a quart jar with marbles in it. A week later try a pint jar and see if pupils estimate more closely, based on their previous estimates.

Activities for Elementary School Mathematics

Page 7.7

WATER, WATER EVERYWHERE

Topic: ESTIMATION

Grade Level: 3-4

Activity Time: 1 class period

Materials Needed:
1. One penny per group of two pupils
2. An eyedropper for each group
3. Water in containers, so each group has its own source and does not have to share containers of water
4. Pencil and paper for each group

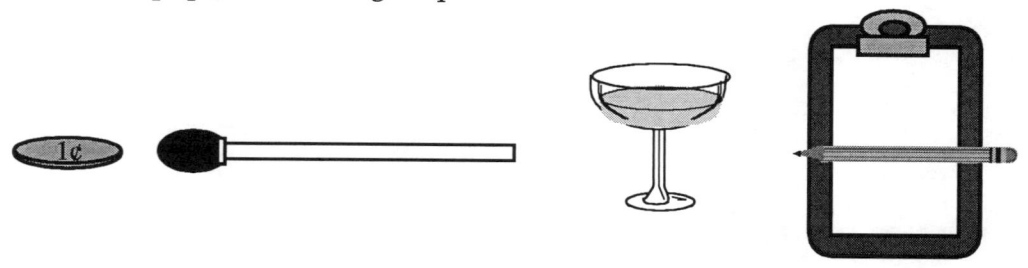

NOTE: pupils may perform this activity singly rather than in pairs, if there are enough materials to go around.

Objectives:

As a result of this activity, the learner will:
- become more proficient at estimation.

Introduction:
1. Divide the pupils into pairs. Ask the class how many drops of water they think a penny can hold.
2. Instruct each pair to come to an agreed-upon estimate and to write it down on paper.

Major Instructional Sequence:
1. In each group, one pupil puts one drop of water at a time onto the "heads" side of the penny with the eyedropper and the other pupil tallies the number of drops.
2. Ask the pupils to estimate again and reverse their roles, so that each has a chance to place drops of water on the penny and do the experiment again.
3. After pupils have done the experiment a second time, have them compare their two estimates to see if they came closer the second time.

Closure or Evaluation:
1. See who holds the class record for placing the greatest number of drops on the

penny without it spilling over.
2. Ask if they think that using the other side of the penny would produce a different result. (Experiment to find out, if time permits.)
3. Ask the class if they think the results would be the same if a new penny was used instead of an older, worn penny. (Experiment to find out, if time permits).
4. Ask if they think smearing a little soap on the face of the coin would have an effect. (Experiment to find out, if time permits. Soap breaks the surface tension of water causing it to immediately drain off the coin.)
5. Seek ways to display your class data.

Activities for Elementary School Mathematics

FRIDAY ESTIMATE

Topic: ESTIMATION

Grade Level: 3-6

Activity Time: 1 class period

Materials Needed:

1. Different items for estimation (such things as cheese puffs, M&M's, or other things that can be counted carefully and perhaps eaten)

cookies in baggie

marbles in cup

large pod of grapes

watermelon seeds

Cheese Puffs

malted milk balls

candy

M & M's

NOTE: When choosing edible items for estimation, you may want to take the neatness factor into account (e.g., M&Ms are less messy than cheese puffs; watermelon seeds may require more extensive cleanup than would grapes).

2. Cards with a pupil's name on each, numbered one through five in each of two columns

Emily	
1.	1.
2.	2.
3.	3.
4.	4.
5.	5.

Page 7.10   Activities for Elementary School Mathematics

Objectives:
As a result of this activity, the learner will:
- develop better estimation skills.

Introduction:
1. On Friday morning, distribute the cards to the class, explaining that they will be used once a week on each Friday for an estimation project. Show the class the item to be estimated that week (for example, a small bag of M&M's) and ask them to estimate how many are in the bag.
2. Instruct the pupils to write their estimates on their cards in the first column, at number 1. Collect the cards and put them on your desk until after lunch (when the count of the item will occur).

Major Instructional Sequence:
1. Sometime after lunch, redistribute the pupil cards and record some of their estimates on the chalkboard.
2. Select two pupils to work with you in counting the items.
3. Have pupils record the exact amount of the count next to their estimates on their cards in the second column, number one.
4. If M&M's or some other edible item was estimated, the class could eat them at this time. (On a later Friday, use a larger bag of M &M's to estimate. Encourage pupils to compare the size of the new bag to the smaller size that was estimated earlier and make a new estimate based on their previous experience.)

9-oz. bag    16-oz. bag    9-oz.    16-oz.

Closure or Evaluation
The pupil (or pupils in case of ties) with the closest estimate may be in charge of distributing the edible item to the rest of the class.

Variation:
1. In the following weeks, incorporate other, not necessarily edible, items into the estimation process. For one week, the amount of marbles in a pint jar could be estimated. A week or two later, the estimation could be for larger marbles in the same size jar or a larger jar containing the regular marbles.
2. Estimations will improve after the pupils have experienced estimates involving similar items or quantities. The range of estimates will narrow as experienced pupils base new tasks on results of their previous experiences.

Activities for Elementary School Mathematics

LENGTH ESTIMATE

Topic: ESTIMATION OF STANDARD LENGTHS

Grade Level: 3-6

Activity Time: 1 class period

Materials Needed:
1. A one-foot ruler, a yardstick, a meter stick, a decimeter length, or any other length-measuring tools to work with in the classroom (These should be available for pupils' use prior to this activity and pupils should be familiar with their use.)
2. Paper and pencil for each pupil

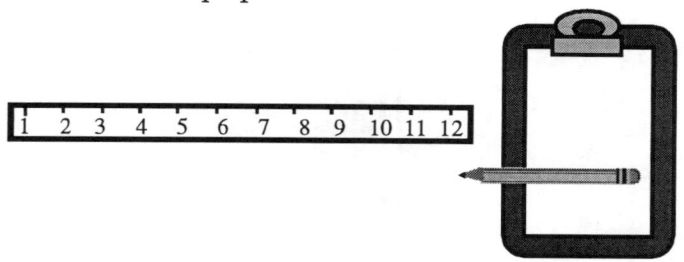

Note: *You must be familiar with any units of measurement that will be discussed in class.*

Objectives:
As a result of this activity, the learner will be able to:
- estimate standard lengths better.
- estimate metric lengths better.

Introduction:
1. Hold up one of the items, such as the foot ruler. (One foot is a good measure with which to begin.)
2. Instruct the pupils to look around for things in the classroom that are approximately the same length as the ruler.

Major Instructional Sequence:
1. Instruct pupils to write down several items that they feel are approximately the length of the ruler you are holding up in front of the class.
2. Ask for a volunteer to get one of the items that they wrote down and bring it to the front of the room. If the pupil chose a book, for example, ask him or her to measure it to see how accurate the guess was.

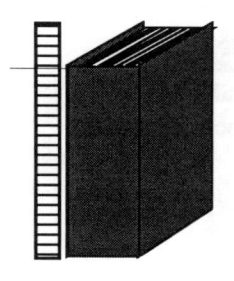

This book is a little short of one foot.

3. If the item the pupil chooses cannot be moved, instruct the pupil to take the ruler to the item and measure it there.

Closure or Evaluation:
1. Continue the activity, using other pupils' suggestions and permitting them to measure the objects with the ruler.
2. Use another measuring tool, such as the meter stick, for the next hunt.

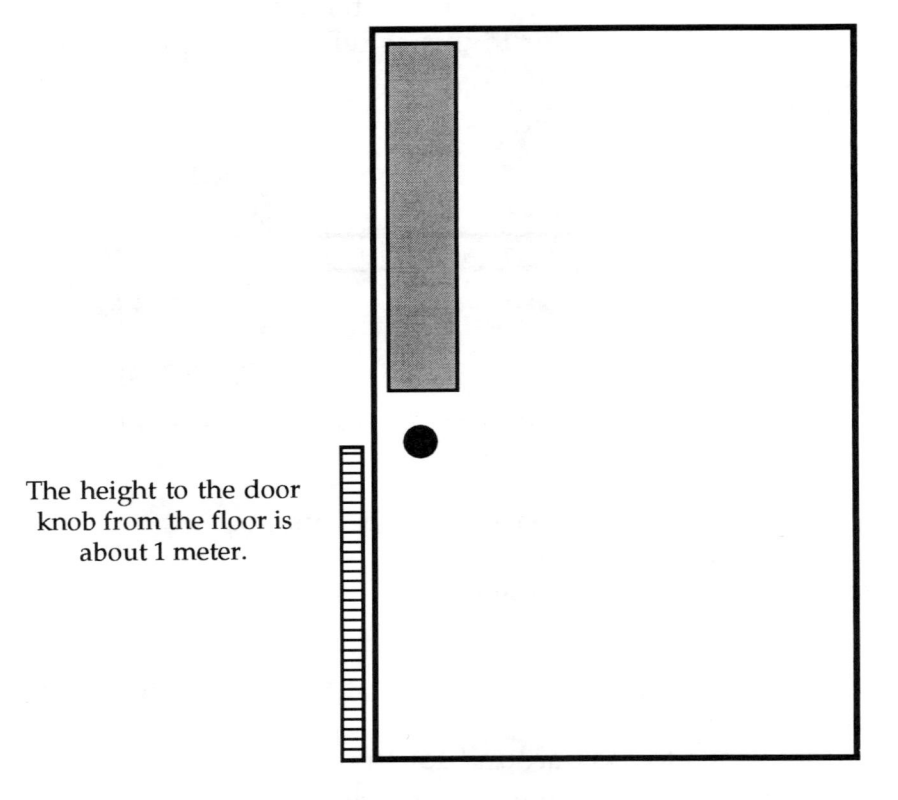

The height to the door knob from the floor is about 1 meter.

If a decimeter were to be used next, the width of an adult hand is approximately a decimeter wide.

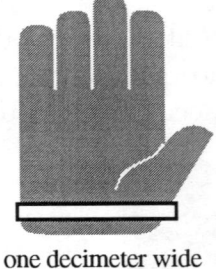

one decimeter wide

Activities for Elementary School Mathematics

GUESSTIMATE

Topic: ESTIMATION

Grade Level: 4-6

Activity Time: 1 class period

Materials Needed:
1. An eight-oz. cup
2. A pint jar or carton
3. A quart jar or carton
4. A half-gallon jug or carton
5. An empty gallon jug or carton
6. Paper and pencil for each pupil
7. Enough marbles to fill the gallon container

Objectives:
As a result of this activity, the learner will:
- estimate numbers.
- make conjectures based on previously learned information.
- become familiar with standard units of measure, such as cup, pint, quart, half-gallon, and gallon.

Introduction:
1. Place the containers side by side on a table where the pupils can look at all the containers at the same time.

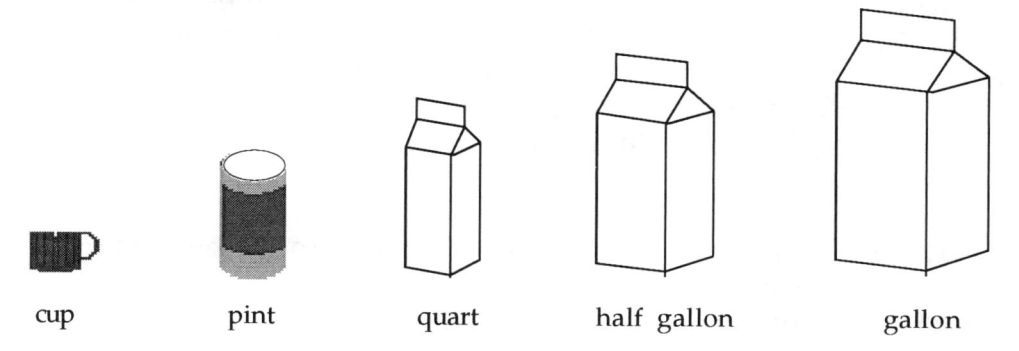

cup   pint   quart   half gallon   gallon

Major Instructional Sequence:
1. Fill the one-cup measure with marbles and ask the class to "guesstimate" how many are in the cup without actually counting the marbles.
2. Instruct the pupils to record their guesstimates. Then let someone count the marbles in the cup.
3. Ask pupils to compare their guesstimates to the actual number of marbles that were in the cup and write the actual number of marbles on their papers.
4. Place the cup measure beside the pint and ask the pupils to compare the two containers. Ask the pupils how much larger they think the pint is than the cup.

Page 7.14   Activities for Elementary School Mathematics

5. Fill both the cup and the pint with marbles. Ask the class to guesstimate how many marbles they think the pint will hold and ask them to write that number on their papers.
6. Select a pupil to count the marbles that were in the pint.
7. Ask pupils to write on their papers the actual number of marbles that were in the pint. Instruct them to compare their guesstimates with the actual number of marbles that were in the pint.
8. Based on the two examples the class has seen, have them estimate how many marbles the quart, half-gallon, and gallon will be able to hold. Place the containers in random order and permit pupils to guesstimate the difference in their sizes.

Closure or Evaluation:
After pupils have recorded their guesstimates, fill the half-gallon with marbles. Ask a volunteer to count the marbles in the half-gallon and explain why twice as many marbles should be needed to fill the gallon container. Ask another volunteer to refill the half-gallon with marbles, pour them into the gallon container, and repeat the procedure (he or she should be able to put two half-gallon containers of marbles into the gallon container).

Variation:
Display three smaller-sized containers that hold the same amount (such as one pint) but have different shapes. Ask pupils to guesstimate what amount the containers hold before you tell the class they are all the same unit of measure. Ask volunteers to fill the containers with marbles and then count the marbles in all three containers to verify the amounts.

pint   pint   pint

Activities for Elementary School Mathematics

Page 7.15

BLINKING EYES

Topic: ESTIMATION

Grade Level: 4-6

Activity Time: 1 class period

Materials Needed:
1. Paper and pencil for each pupil
2. A stopwatch, alarm clock, wall clock, or a watch with a second hand (You must time this activity for thirty seconds.)
3. Calculators for each pair of pupils

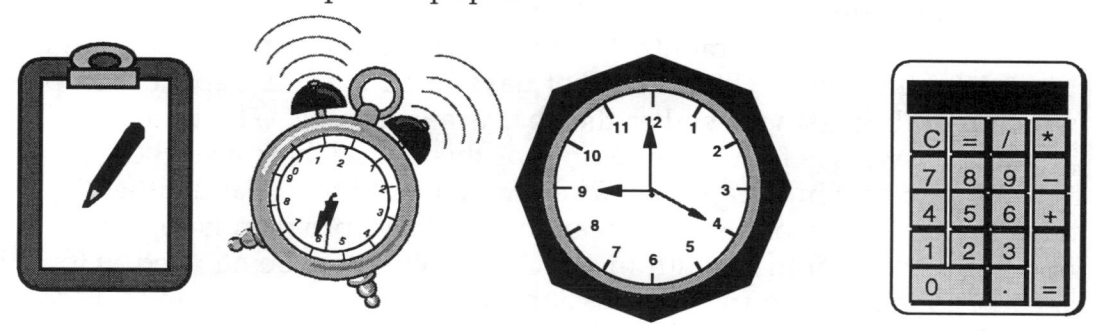

Objectives:
As a result of this activity, the learner will:
- use estimation skills.
- use a calculator to help solve problems.

Introduction:
1. Separate pupils into groups of two.
2. On your signal, you will ask pupils to begin counting the number of times that their partner blinks his or her eyes. Do not reveal how long they are going to be timed, only to be as natural as possible while being observed.

Major Instructional Sequence:
1. Give the signal for pupils to begin counting their partners' eye blinks.
2. Time the pupils for thirty seconds.
3. Tell "counters" to record the number of blinks on their papers.
4. After the blink count is recorded, tell the pupils that they were timed for thirty seconds.
5. The first question to ask is, "How many blinks would have occurred if you had been timed for one minute?" (Pupils may use a calculator or mentally multiply by two.)
6. The second question: "How many blinks would there have been in one hour?" (Pupils must multiply by sixty this time; they may use calculators.)

7. The next question: "How many blinks in a day?" (times 24) The idea of how much sleep is received in a normal 24-hour period must be addressed at this point. One way to address this element is to subtract from 24 the number of hours of sleep the pupil claims to have had and multiply that number times the hourly blink rate.

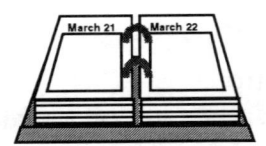

8. "How many blinks in a week?" (times 7)
9. "How many blinks per month?" This poses an interesting and thought-provoking question. The number of days in a month is dependent upon which month it is and whether the year is a leap year. The number of possible answers is four. The reason for this is that there are 28 days in February (except in a leap year, when there would be 29) and either 30 or 31 days in the other months. One way to avoid any problem is to use the number of days in the month in which the activity is being used. However, it is a good idea to encourage discussion to take place so that pupils discover that there is more than one possible answer.
10. "Number of blinks in a year?" is another thought provoking question for pupils to answer. (Normal years have 365 days and leap years contain 366 days. The exact number of days actually calculate as 365.25 days.)

Closure or Evaluation:
1. Encourage pupils to consider and discuss other factors that will affect their answers, such as smoky air, dry air, dry eyes, contact lenses, and any other conditions that may affect the number of blinks of their partners' eyes.
2. The total number of blinks, for a pupil that blinked his or her eyes three times during the thirty-second period, would be 2,102,400 (that is, if the pupil slept an average of 8 hours a day and the year is not a leap year).
3. Answers that are way above or way below the correct number could be checked to find any mistakes that may have occurred.

Activities for Elementary School Mathematics   Page 7.17

RIGHT PRICE ESTIMATE

Topic: ESTIMATION

Grade Level: 4-6   **Activity Time:** 1 class period

Materials Needed:
1. Catalogs or advertising brochures with pictures of items for sale
2. Poster boards (one for each collection of advertised items)
3. Paper, pencil, and calculator for each pupil

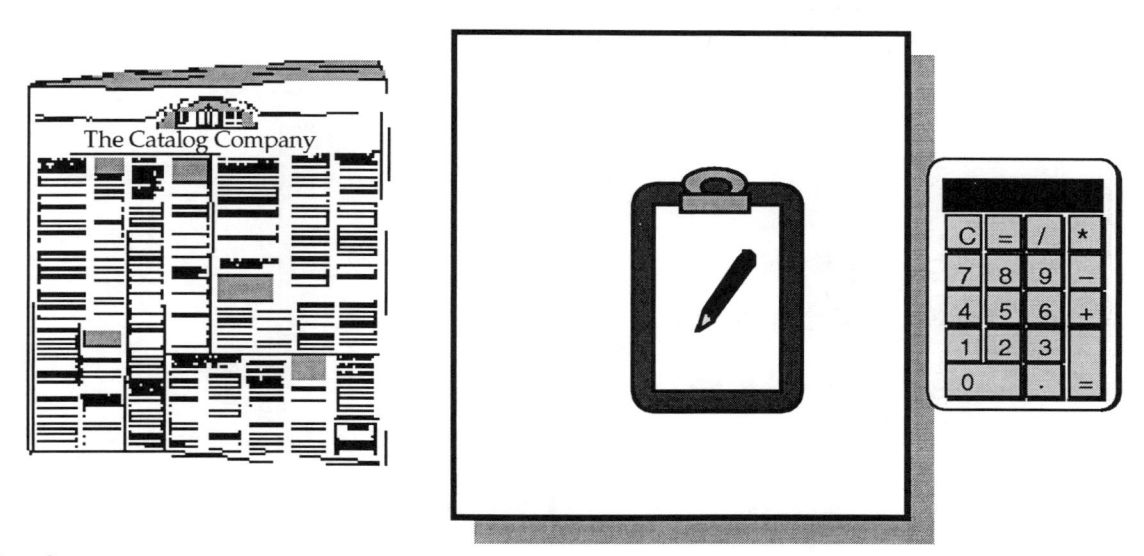

Objectives:
As a result of this activity, the learner will:
- improve calculator and estimation skills.

Introduction:
1. Before class, choose the items from the catalogues/brochures to be used for the right price game and arrange them on a piece of poster board (including the prices).

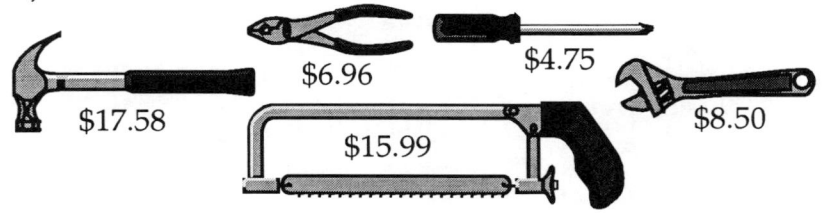

$6.96 $4.75
$17.58 $8.50
$15.99

In this example, pictures from a hardware store's advertising brochure have been used.

2. Cover the prices with pieces of construction paper, attached by a small piece of tape so that the paper may be easily removed later in the game.

Major Instructional Sequence:
1. Ask the class to write on their papers their estimates for the price of each item.
2. When the class is finished estimating, remove the construction paper to reveal the prices. Pupils will enter the actual price of an item in their calculators and subtract their estimates from that price.
3. If a pupil overpriced the item, his or her answer will show as a negative number on the calculator.

Closure or Evaluation:
The winner per item is the pupil whose estimate comes closest to the actual price without going over the cost of the item.

Activities for Elementary School Mathematics

ESTIMATE/MEASURE IT

Topic: ESTIMATION AND MEASUREMENT

Grade Level: 4-6

Activity Time: 1 class period

Materials Needed:
1. 3 x 5 index cards (one card per pupil), each with measurements on them appropriate for the grade level (such as one foot, six inches, 10 cm, etc.)
2. Rulers or tape measures
3. Paper and pencil for each pupil

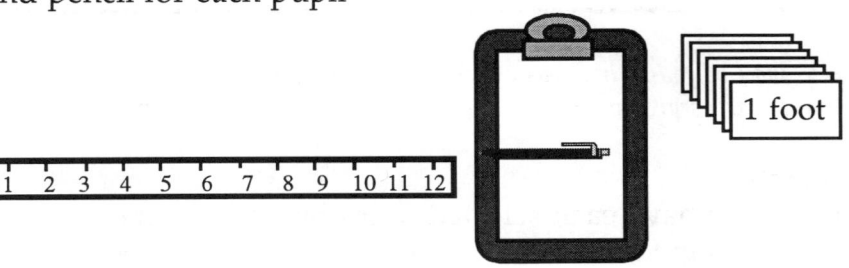

Objectives:
As a result of this activity, the learner will:
- estimate the length of objects to the nearest foot or inch (or whatever measurement is being emphasized in that lesson).
- measure the lengths of selected objects in the lesson.

Introduction:
1. Before class, prepare the index cards, writing one measurement on each of them.
2. Shuffle the cards and distribute a card to each pupil.
3. Ask the pupils to copy the unit of measure from their cards onto the left side of their papers. Explain that this measurement will be the unit of measure that each pupil will use to "guesstimate" measurements of items in the classroom.

length	item	actual length
1 foot	math book	
	computer screen	
	unsharpened pencil	
	magazine	

Major Instructional Sequence:
1. Tell the class that they should search for items that they believe are

approximately the same length as the measurement written on their cards. If his or her card reads "one foot," for example, a pupil will search for four items which he or she thinks are one foot long.
2. Instruct the class to list the objects on paper *before* they measure them.

length	item	actual length
1 foot	math book	11 inches
	computer screen	10 inches
	unsharpened pencil	8 inches
	magazine	1 foot

NOTE: *Items also may be found in areas around the school and playground, depending on what length the pupils are searching for and how long they have to complete the activity.*

Closure or Evaluation:
After the pupils have each selected four objects, they measure each item with a ruler or tape measure to see how accurate their estimates were. Encourage discussion of the results, asking the pupils to share their choices of items and lengths of measure with the class. Pupils will quickly learn to measure objects and estimate their lengths accurately.

Activities for Elementary School Mathematics | Page 7.21

OH, HOW LONG, HOW WIDE, HOW HIGH

Topic: ESTIMATION AND MEASUREMENT

Grade Level: 4-6   **Activity Time:** 1 class period

Materials Needed:
1. Paper and pencil for each pupil
2. Ruler or tape measure for each pupil
3. For each pupil, a list of objects to be estimated and measured (reproduced on a copy machine)

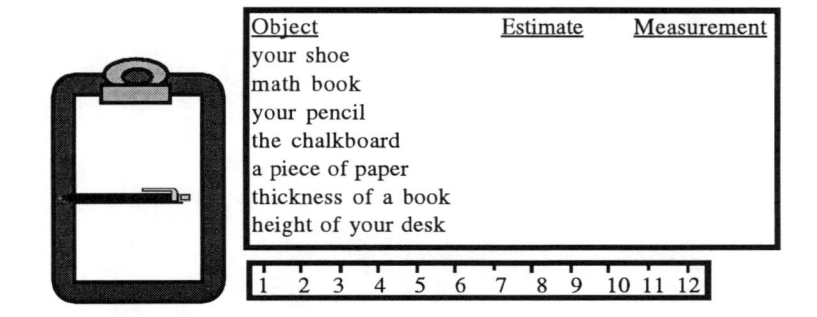

Object	Estimate	Measurement
your shoe		
math book		
your pencil		
the chalkboard		
a piece of paper		
thickness of a book		
height of your desk		

Objectives:
As a result of this activity, the learner will:
- estimate measurements of objects.
- measure objects to nearest quarter inch.

Introduction:
1. Give each pupil a copy of the list of objects.
2. Instruct the pupils to estimate the length, width, or height of each item on the list and record the estimate in the appropriate column on the list.

Major Instructional Sequence:
1. After the pupils have estimated the measurements of the items on their list, they are to measure each item and record the answer on their list in the appropriate column.
2. Ask pupils to compare their estimates with the actual measurements of the objects on the list.

Closure or Evaluation:
If a pupil's estimates are within half an inch of the actual measurement, tell the pupil that he or she did a good job of estimating!

Activities for Elementary School Mathematics

CHAPTER EIGHT
CALCULATIONS

When calculators first appeared on the market, a fiery debate began: Should calculators be used in the classroom? It was said that children would become dependent on the calculator and not use their own abilities to solve problems or learn their basic facts. Calculator use in the classroom is no longer questioned. Now the argument is about *how much* or *when* they should be used.

It remains true that **basic facts should be mastered independently of the calculator**. However, the calculator can be used judiciously to aid in the learning of basic facts, as in the case of a quick check of combinations that have been causing individual pupils difficulty.

During regular mathematics class time, the calculator should be used to verify answers or to solve problems that otherwise may not be completed. The use of the calculator in problem solving should be a major consideration. It can free pupils to be creative and to explore ideas of which they otherwise may not be capable.

Modern calculators can perform an amazing variety of operations. There are calculators that can add, subtract, multiply, and divide fractions.

In this display, the 6 is a whole number and 1/2 makes it a mixed number.

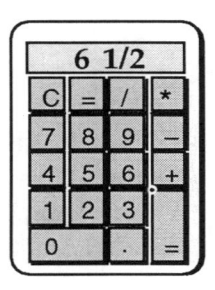

The calculator can work with mixed numbers and change them into decimal fractions, as in this display.

Fractions can be simplified: pupils can select the greatest common factor to simplify, or they can let the calculator select the common factor. Even if the pupil selects an incorrect GCF, the calculator can handle that by not using it or by simplifying as far as it can and indicating that it is not completely simplified.

Page 8.2   Activities for Elementary School Mathematics

Decimals can be rounded to a desired place: hit "FIX" 3 to round 4.56482 to the nearest thousandths (3 places after the decimal point) and the answer is rounded to 4.565.

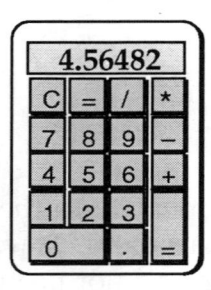

Incorrect entries can be backspaced out and corrected. Constant operations can be performed. The list goes on and on.

Activities have been included in this section to help familiarize the pupils with many of the operations modern calculators can perform.

Activities for Elementary School Mathematics

CALCULATOR BASEBALL

Topic: USE OF A CALCULATOR/MENTAL MATHEMATICS

Grade Level: 1-6   **Activity Time:** 1 class period

Materials Needed:
1. Two calculators
2. Four bases for a baseball diamond (The bases could be represented by carpet squares, books [not to be stood *on*, only stood *by*], or any other object on hand in the classroom. Bases should be placed on the floor so players can walk around as the game is played.)

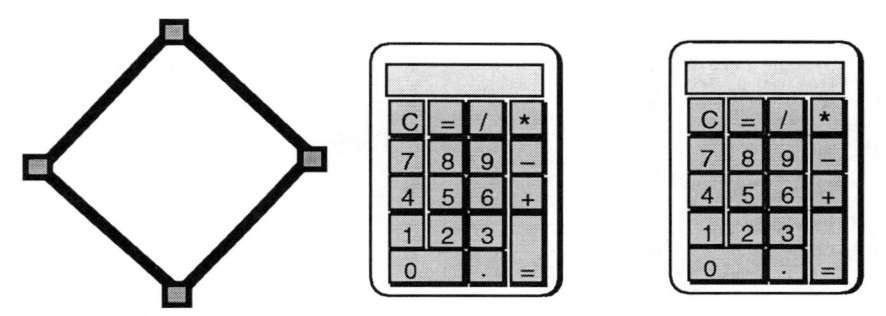

Tape can be placed on the floor to represent the base lines on the baseball diamond, or the game could be played outside on the playground baseball diamond if the school has one.

3. A set of questions appropriate for the age group playing the game (The game is good as a review before a chapter test.)

Objectives:
As a result of this activity, the learner will:
- become familiar with the use of a calculator.
- review previously learned mathematics content.

Introduction:
1. Divide the class into two teams and position them on different sides of the room. Teams may select team names, if they wish.
2. Place the four bases on the floor in front of the room so a baseball diamond can be envisioned. Tell the teams that the players will advance from one base to another as they correctly respond to the questions.

Major Instructional Sequence:
1. Team 1 sends up first batter to home plate.
2. Act as the pitcher for both teams. Ask the first batter to solve a mental math problem.

Page 8.4   Activities for Elementary School Mathematics

3. The second batter for team 1 has one calculator, and first person for team 2 has the other calculator in order to check the first batter's answer.
4. If the pupil answers correctly, he or she goes to first base. If the answer is not correct, the pupil goes to the back of the batting roster and an out is recorded. After three outs, team 2 goes to bat.
5. A scoreboard on the chalkboard, if played inside, can be used to keep track of hits and runs.

Closure or Evaluation:
1. Each correct answer is worth a base. Runners advance until a runner scores.
2. The game continues for a predetermined number of innings. The team with most runs wins. Tie games are decided by determining the team that had the most hits in the game.

	A	B	C	D	E	F	G	H	I	J	K	L	M
1	Inning		1	2	3	4	5	6	7	8	9	H	R
2	Team	Hits	1	2	4	4	0	0				11	
3	1	Runs	0	0	1	1	0	0					2
4													
5	Team	Hits	4	0	0	3	4	1				12	
6	2	Runs	1	0	0	0	1	0					2

After six innings these teams are tied at 2 runs each. If the game ends with both teams having scored the same number of runs, the team with the most hits is the winner. If both teams also have the same number of hits, the game ends in a tie.

Activities for Elementary School Mathematics   Page 8.5

INVERSE OPERATIONS

Topic: USING THE CALCULATOR TO CHECK PROBLEMS

Grade Level: 1-6   **Activity Time:** 1 class period

Materials Needed:
1. A calculator for every two pupils
2. Paper and pencil for each pupil

3. Up to 10 problems which can be solved and checked by using the inverse operation on a calculator

Objectives:
As a result of this activity, the learner will:
- use the calculator to check problems for accuracy.
- understand the inverse operation and how to use it to check problems.

Introduction:
Whatever the grade level and whatever the operation, pupils should be taught that checking with the inverse operation can be accomplished on the calculator. It is a typical method of checking pupils' computational skills.

Major Instructional Sequence:
1. Instruct pupils to work up to 10 problems of the same problem type. For example, first graders can work problems which use the addition facts they have been studying, and they can check them using the inverse operation (subtraction). Explain to them that, using the calculator, 6 + 3 = 9 can be checked by entering 9. Then, using the opposite operation of addition subtraction, take away 3 from 9. This would leave 6.
2. At a higher grade level, if a sixth-grade pupil works a multiplication problem like 761 x 58, it can be checked by division. If the pupil works the problem properly and gets an answer of 44,138, he or she can check by

entering 44,138 into the calculator and dividing by 58.

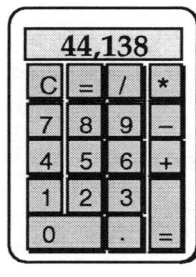

| 761 × 58 | 44,138 ÷ 58 |

3. Astute pupils will realize their answer should be 761 with no remainder. If they have an incorrect answer, they will not get 761 or they will have a remainder (which indicates that their answer is incorrect).
4. As another example, when using the inverse operation to check division problems (for example, 4,891 divided by 53 is equal to 92 with a remainder of 15), a pupil must take the following steps: Multiply the divisor (53) by the quotient (92) and then add in the remainder (15). If all goes well, the correct display on the calculator will be 4,891 (which is the correct answer).

Closure or Evaluation:

After the pupils have become comfortable with the inverse operation, reinforce in them the idea that it is one of the easiest methods by which to check all their computations.

Activities for Elementary School Mathematics

FACTS REVIEW

Topic: BASIC FACTS

Grade Level: 2-5

Activity Time: 1 class period

Materials Needed:
1. Paper and pencil for each pupil
2. A calculator for each pupil (Two pupils may share a calculator for this activity.)

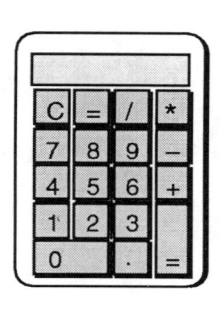

Objectives:
As a result of this activity, the learner will become:
- more proficient with basic facts.
- more familiar with the basic use of a calculator.

Introduction:
Explain to the class that each pupil will use the calculator to solve basic facts and record his or her basic facts answers on paper.

Major Instructional Sequence:
1. Call out a basic fact to the class (3 + 4; 7 - 3; 12 + 5; 3 x 2; 3 x 7; for example).
2. Each pupil is to enter the fact into the calculator, but **instruct them not to press the equals key**. (If two pupils are working on a calculator, they take turns entering the facts you call out.) Circulate among the pupils, checking for understanding and giving assistance as necessary (and spot-checking to make sure pupils are following the rules).
3. Instruct the pupils to write their responses to the fact on their papers.

Closure or Evaluation:
1. When everyone has written their answers, have them press the equals key to check their responses.
2. Play rounds of ten or twenty facts and see who has the most correct.
3. Keep track of pupils, if desired, to see whether individuals are having difficulty with the facts being worked on that day.

Page 8.8   Activities for Elementary School Mathematics

CALCULATOR +

Topic: CALCULATOR SKILLS/BASIC ADDITION

Grade Level: 3-5   **Activity Time:** 1 class period

Materials Needed:
   A calculator for each group of 6 or 8 pupils

Objectives:
   As a result of this activity, the learner will:
   - become more familiar with the use of a calculator.
   - practice mental mathematics in addition.

Introduction:
   1. Group the pupils in a row or a circle, with about six or eight pupils in each group.
   2. Instruct the first pupil in each group to take the calculator and enter a problem (such as 6 + 8, for example), telling pupils beforehand not to press the = sign.

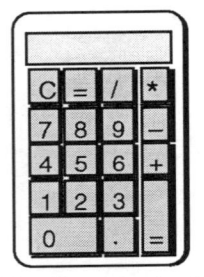

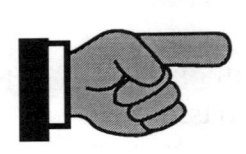

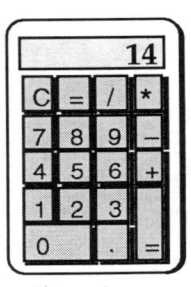

Pupil enters 6 + 8.   Hands calculator to next pupil to give answer **before** pressing the "=" key to check.

Major Instructional Sequence:
   1. The calculator is passed to the second person in each group, and this pupil must give the answer and then press the = key to see if his or her answer was correct.
   2. The second pupil then uses his or her answer to the first problem as an addend for the next problem, entering 14 + a single-digit number (such as 5). The calculator is passed to the next person in the line.

Closure or Evaluation:
   1. The calculators continue to circulate around the groups.
   2. If a pupil's response is wrong, he or she drops out of the group.
   3. The winner from each group is the last pupil left in the group, having answered all correctly.

Activities for Elementary School Mathematics

CALCULATOR–

Topic: CALCULATOR SKILLS/BASIC SUBTRACTION

Grade Level: 3-5 **Activity Time:** 1 class period

Materials Needed:
A calculator for each group of 6 or 8 pupils

Objectives:
As a result of this activity, the learner will:
- become more familiar with the use of a calculator.
- practice mental mathematics in subtraction.

Introduction:
1. Group the pupils in a row or a circle, with about six or eight pupils in each group. Provide a calculator to the first pupil in each group.
2. Tell the first pupils to enter 100 on their calculators and then enter a single-digit number to subtract from the 100. Remind pupils beforehand not to enter the = sign.
3. The first pupil enters 100 - 8, for example.

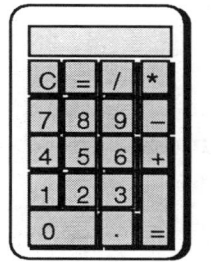

Pupil enters 100 – 8. 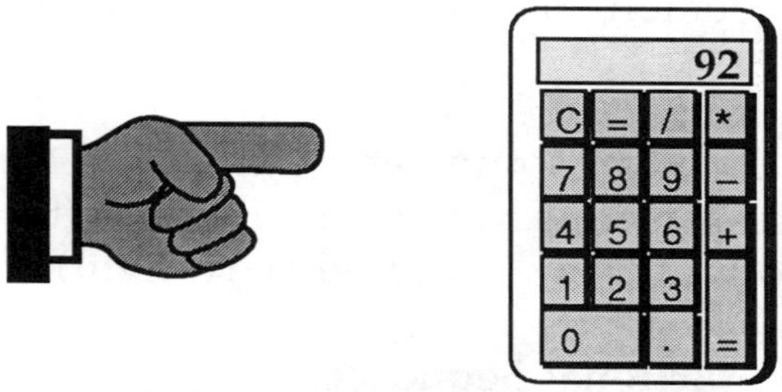   Hands calculator to next pupil to give answer **before** pressing the "=" key to check.

Major Instructional Sequence:
1. Instruct the pupils to pass the calculators to the second person, and this pupil must say the answer and then press the = key to see if his or her answer was correct.
2. The second pupil then uses his/her answer to the first problem as an addend for the next problem, entering 92 – a single-digit number (such as 5). The calculator is then passed to the next person in the line for an answer.

Closure or Evaluation:
1. The calculator continues to circulate around the groups.
2. If a pupil response is wrong, he or she drops out of the group.
3. The winner from each group is the last pupil left in the group, having answered all correctly.

Activities for Elementary School Mathematics

WIPE-OUT

Topic: CALCULATOR SKILLS/PLACE VALUE

Grade Level: 3-6   **Activity Time:** 1 class period

Materials Needed:
   A calculator for each pupil

Objectives:
   As a result of this activity, the learner will:
   - become familiar with a calculator.
   - be able to read a large number.
   - demonstrate an understanding of place value in a large number.

Introduction:
   1. Write a large number on the chalkboard (such as 3,452,168).
   2. Ask someone to read the number (three million, four hundred fifty-two thousand, one hundred sixty-eight).

Major Instructional Sequence:
   1. Ask the pupils to enter the number (3,452,168) on their calculators.
   2. Tell the class you want them to "wipe out" the five in the number. Do not tell them that the five is actually fifty thousand.
   3. The pupils must subtract fifty thousand from the number on their calculators in order to remove the five from the display.
   4. Circulate among the pupils, checking for understanding and giving assistance as necessary. Hopefully, all pupils have the correct answer on their calculators (3,402,168).
   5. Repeat the process (for example, this time ask the pupils to wipe out the one, which is in the hundreds column).

Closure or Evaluation:
   1. Encourage discussion about the calculator action that was necessary in order to wipe out the requested digits.

Activities for Elementary School Mathematics   Page 8.11

2. Pupils must know the value of the digit (in which column it is located) in order to wipe it out.

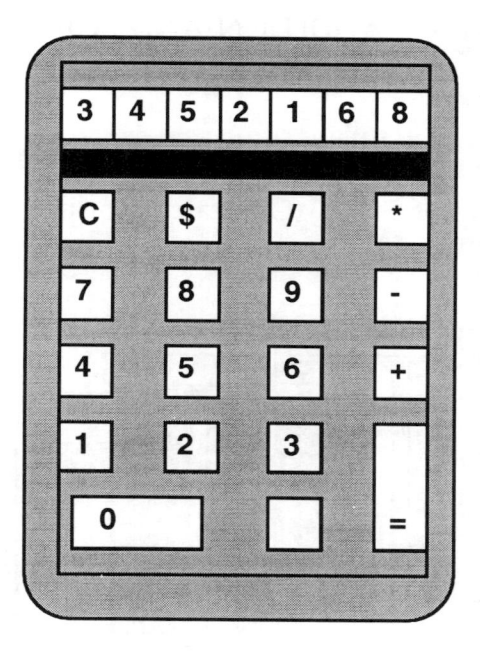

Variation:
Pupils in the higher grades could play the same game with a decimal number. For example, 3452.168 could be used, and you would present "wipe-out" numbers with tenths, hundredths, and thousandths in them. The game would remain the same with the exception that pupils must understand decimal columns in order to wipe out any number to the right of the decimal point.

Page 8.12   Activities for Elementary School Mathematics

TARGET-50

Topic: CALCULATOR SKILLS/ADDITION AND SUBTRACTION

Grade Level: 4-6   **Activity Time:** 1 class period

Materials Needed:
One calculator per group of 5 or 6 (One calculator per row, works well, if classroom is aligned in rows.)

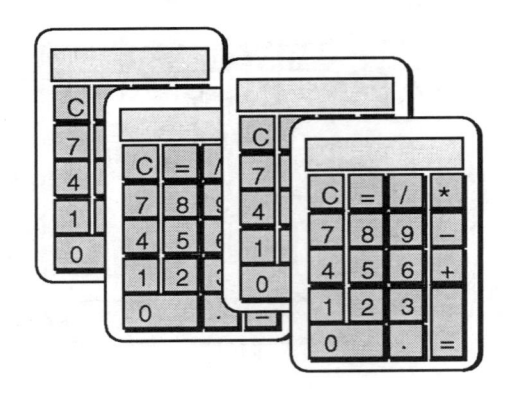

Objectives:
As a result of this activity, the learner will:
- gain skill in estimation.
- gain skill in addition and subtraction.
- develop a strategy for reaching a specific target number.

Introduction:
1. Select a target number (90, for example) and instruct the first pupil in each group to enter a two-digit number on their calculators (other than the target number itself) that is less than the target number.
2. Instruct the pupils to perform mental calculations to determine the other number which would cause the sum to be 90 and then add that number to the first one on their calculators.

Major Instructional Sequence:
1. Before the first pupils enter their numbers of choice into their calculators, make sure that the calculators are at zero.
2. After the first pupils correctly perform their calculations, tell them to add another number to the target number. The calculators are then passed to the next pupil in each group. These pupils must perform mental *subtraction* in order to determine the other number which would cause the *remainder* to be 90.
3. After the second pupils correctly perform their calculations, tell them to

subtract another number from the target number. The calculators are then passed to the next pupil in each group. This will take the cycle back to its beginning, where pupils must mentally determine the other number needed to bring the total to 90, and the game continues.
4. The calculator keeps going around the group or down the row, with each pupil reaching the target number.

Closure or Evaluation:
A winner is any pupil who enters a number and brings the sum (or remainder) on the calculator to the target number.

Page 8.14   Activities for Elementary School Mathematics

YOUR ORDER, PLEASE

Topic: CALCULATOR SKILLS/PERCENTAGES

Grade Level: 4-6   **Activity Time:** 1 class period

Materials Needed:
  1. Menus from several local restaurants

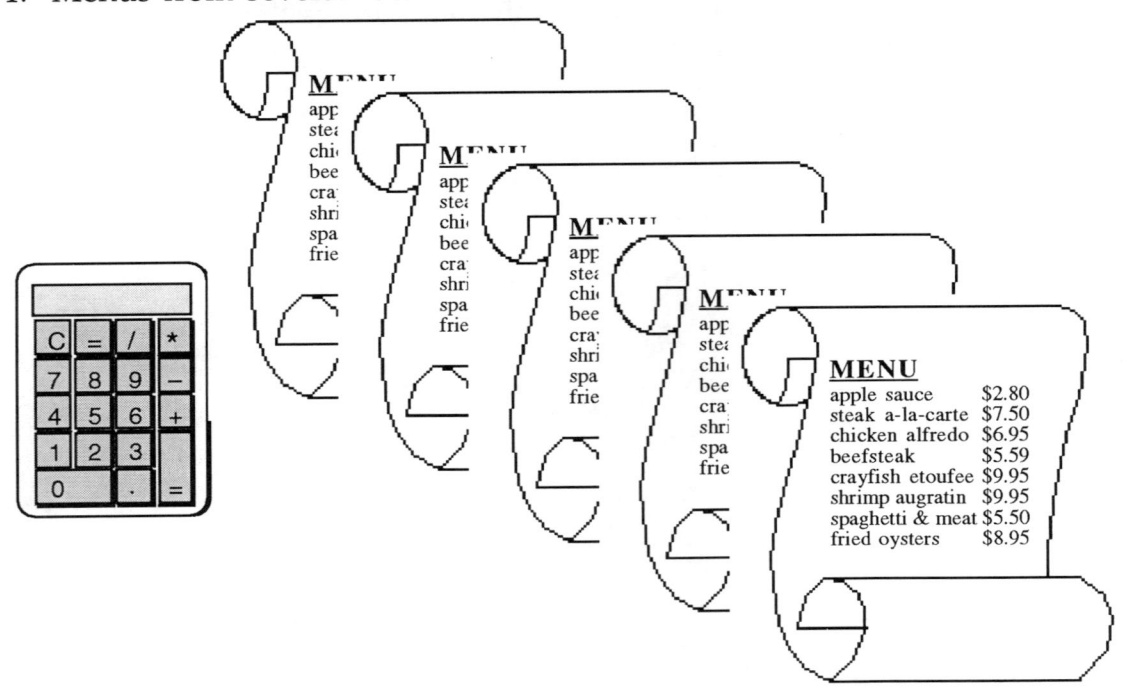

  2. A calculator for each pupil (This activity may also be performed by groups of two or three pupils, with a calculator for each group.)
  3. Paper and pencil for each pupil

Objectives:
  As a result of this activity, the learner will:
  - become familiar with a calculator.
  - use a calculator to figure percentages (for tips or discounts).

Introduction:
  1. Tell the class to imagine that they are going to a restaurant and must select what they want to have for a meal, including a beverage, appetizer, and dessert.
  2. Pass out a menu to each pupil.
  3. Set a target amount of $15 to $20 (higher or lower, depending on the prices on the menus) that the pupils will aim for (excluding tip).

Activities for Elementary School Mathematics

4. Select certain meals and inform the class that these meals are daily "specials" which offer a 25% discount from the listed price.

Major Instructional Sequence:
1. Instruct the pupils to select items from the menu, trying to stay under the target amount. The object of the game is to see which pupil can select items that will get him or her closer to the target amount than any of the others.
2. The pupils will add up different food, drink, dessert, and appetizer combinations. If a pupil goes over his or her goal, items may be substituted and the cost refigured.

Closure or Evaluation:
The pupil who gets closest to the target amount wins.

Variations:
1. If you desire to make this activity more challenging, tell the class that a 15-20% tip is usually expected in most restaurants. The tip could then be figured into the pupils' calculations, but set a percentage for everyone to use and instruct the pupils to add this amount to the meal price. Pupils should use a calculator to get the total cost of the meal and tip.
2. Tell pupils to select meals for a family of four or five, giving them a new target amount suitable for a family.
3. Play money may be used in this exercise, giving pupils experience with counting out amounts of money and figuring change.

Activities for Elementary School Mathematics

CHAPTER NINE
MISCELLANEOUS MATHEMATICAL CONCEPTS

This section contains different topics, and no two are exactly alike. Two of the activities deal with *money*. The value of various combinations of coins is discussed in one activity so children can learn that there is more than one way to show an amount of money. Another activity deals with a calendar and represents each day as an amount of money. A second approach to the calendar could be to place a straw in a container for each day and when the straws reach ten they are bundled and put into the tens container. Some teachers have a Hundred Day celebration when ten groups of ten is accomplished for the hundredth day of school. They do activities based on the number 100.

One activity, Bingo Times, covers *time* and addresses times written in different ways. 11:55 is also 5 minutes till 12. A clock face also shows the time.

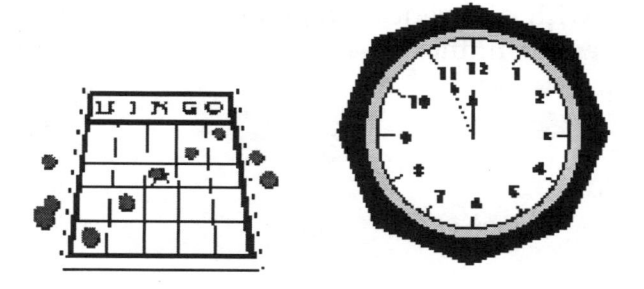

Another activity covers *integers*, addressing negative and positive numbers. Dice are rolled and pupils record their golf score, which could be a negative number in the game.

Statistics is covered with the use of sports cards and mythical test scores. Statistics of sports stars and of a mythical class's test scores are used to generate interest. Finding batting averages, passing percentages, or winning percentages are a few of the statistical activities that can be explored with sports cards. Constructing a frequency distribution and computing the median, mode, and mean, and exploring various percentages are explorations that can be done with test scores.

Probability is also addressed, using a newspaper to find out which letter occurs most often. In this activity several selections are circled and the pupils must find the most often-used letters. They could predict in advance which letter will appear most in a randomly selected passage in the newspaper. Another activity addressing probability involves the use of a spinner.

Other activities deal with different mathematical concepts, such as reading large numbers, different names for numbers, and using the bar graph.

Activities for Elementary School Mathematics

DAYS ARE MONEY

Topic: MONEY

Grade Level: 1-2

Activity Time: 1 class period

Materials Needed:
1. A large calendar with plenty of room on the space allotted to each day

Sunday	Monday	Tuesday	Wednesday	Thursday	Friday	Saturday
			1	2	3	4
5	6	7	8	9	10	11
12	13	14	15	16	17	18
19	20	21	22	23	24	25
26	27	28	29	30	31	

2. Coins (or representations of coins/play money) in the amount of penny, nickel, dime, and quarter (Each child will need at least 61 pennies, 12 nickels, 20 dimes and 7 quarters.)
3. Worksheets with the current month's calendar on them for each pupil

Objectives:
As a result of this activity, the learner will:
- become more familiar with coins and their values.
- add coins to total thirty-one.

Introduction:
1. First, distribute the calendar sheets to the pupils.
2. Distribute the coins to the children and explain that they should determine the monetary equivalent of the date on each space of the calendar sheet.

Page 9.4   Activities for Elementary School Mathematics

Major Instructional Sequence:
1. Instruct the children to determine the fewest number of coins needed to represent each day of the month. (The 12th, for example, would be one dime and two pennies.)
2. Ask the pupils to glue or tape the coin representations on the corresponding days on the calendar.
3. Since it is not a race, children may work together if they need help.

Closure or Evaluation:
1. Circulate among the pupils, checking for understanding and giving assistance as necessary.
2. Encourage class discussion about the proper answers, as well as other correct amounts which do not the use fewest coins required.

Variation:
A calendar sheet could be created by the entire class as a poster for classroom display.

Sunday	Monday	Tuesday	Wednesday	Thursday	Friday	Saturday
			1 (1¢)	2	3	4
5	6 (5¢)(1¢)	7	8	9	10 (10¢)	11
12	13	14	15	16	17	18
19	20 (10¢)(10¢)	21	22	23	24	25 (25¢)
26	27	28	29	30	31 (25¢)(5¢)(1¢)	

Activities for Elementary School Mathematics   Page 9.5

BINGO TIMES

Topic: TELLING TIME

Grade Level: 1-3   **Activity Time:** 1 class period

Materials Needed:
1. Pre-made Bingo Times cards for everyone in the class

T	I	M	E	S
🕐	11:30	🕐	10:15	🕐
12:00	🕐	7:30	🕐	8:45
🕐	6:00	🕐	9:45	4:30
1:30	🕐	11:15	3:30	11:00
6:45	10:30	🕐	5:15	🕐

   NOTE: *The clock faces on the cards should have all the numerals on them if intended for use by younger children.*

2. Markers (such as beans or buttons)
3. A master list of times appropriate for the grade level
4. Three sets of 3 x 5 index cards with the letters **T I M E S** (one letter per card, providing 15 cards; 3 Ts, 3 Is, 3 Ms, 3 Es, and 3 S cards)

Objectives:
   As a result of this activity, the learner will:
   • equate what is on a clock face with verbal and written expressions of time.

Introduction:
1. Distribute the materials to the pupils.
2. Tell the class that this game is played like BINGO but uses the word TIMES at the top of the card instead of BINGO.

Major Instructional Sequence:
1. Shuffle the index cards and place them in a stack.
2. Read out a letter/time combination, selecting the letter from the card on top of the stack (then place the card on the bottom of the stack), and using the master list to select a time. For example, the combination could result in you calling, "T, 3 o'clock." (Be sure to keep a master list of the combinations called for later checking.)
3. If a pupil has a digital reading of 3:00 or a clock face which represents three o'clock, in the T column, he or she may place a marker on the appropriate space on the card.
4. Call out the next letter/time combination, and the game continues.

Closure or Evaluation:
1. Play continues until a pupil who has covered five across, down or diagonally calls out, "TIMES!"
2. Cards should not be cleared until you check to verify the win.

Activities for Elementary School Mathematics   Page 9.7

BAR GRAPH

Topic: BAR GRAPH CONSTRUCTION and READING

Grade Level: 1-6   **Activity Time:** 1 class period

Materials Needed:
1. Double-sided sticky tape (enough for 4 long pieces to put on a poster board for display in the front of the room)
2. 2 x 2 squares of construction paper, 4 colors, approximately 20 of each (depending on class size)
3. Pictures of 4 desserts

cookies   ice cream cone   ice cream bar   cake

4. 4 brown paper bags with a picture of one dessert taped on each of them

Objectives:
As a result of this activity, the learner will:
- construct a simple bar graph.
- read a bar graph.

Introduction:
1. Before class, prepare a poster board by placing the double-sided tape vertically to represent a column under a picture of each dessert (see illustration below).

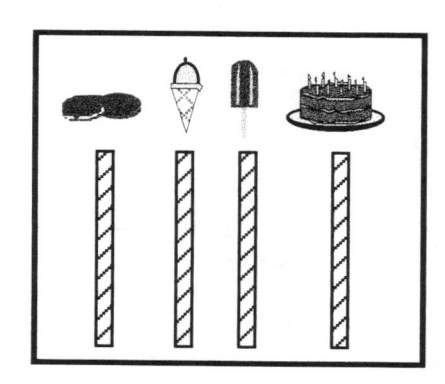

Page 9.8   Activities for Elementary School Mathematics

2. During class, place one set of colored squares in each bag. Tell the pupils that they are going to determine which of the four desserts is best liked by the class.
3. Tell the pupils they must select one square, only from the dessert bag representing the selection they like best.
4. Choose four pupils, each to carry a bag and walk around the room so each pupil can select one dessert.

Major Instructional Sequence:
1. Display the poster board, perhaps by propping it on the chalk tray at the front of the room.
2. Ask pupils to come up in small groups (about 4 or 5 pupils at a time) to put their squares on the tape under their favorite dessert. Tell them to carefully line up their squares as neatly as possible, one under the other, so that an accurate picture will emerge.

Closure or Evaluation:
Depending on the grade level, discussion can be encouraged by asking different questions about the resulting information on the bar graph. (Answers to the sample questions represent the bar graph results illustrated above.)

A sample of lower grade questions:
1. Which dessert did the most people select as their favorite? (the ice cream cone)
2. How many pupils selected it as their favorite? (6)
3. Which was the least chosen dessert? (the ice cream bar)
4. How many chose this dessert? (2)
5. How many more pupils chose the ice cream cone than the ice cream bar? (4)
6. Which two desserts were selected by the same number of pupils as their favorite? (the cookies and the cake)

Activities for Elementary School Mathematics

A sample of upper grade questions:
1. What was the mean score of the data above? (4 — add scores together and divide by the number of scores)
2. What was the median for the data? (4 — the middle score when scores are arranged in order; known as a frequency distribution)
3. What was the mode? (4 — the score that appears most often)
4. What was the range of the data? (1-6 — low to high score. Some books would put the range as 5. Five is the difference in the high and low scores.)

Page 9.10   Activities for Elementary School Mathematics

I'M FOR TWO

Topic: MONEY

Grade Level: 2-3   **Activity Time:** 1 class period

Materials Needed:
1. 8 x 10 picture cards with a picture and price of a different item on each of them

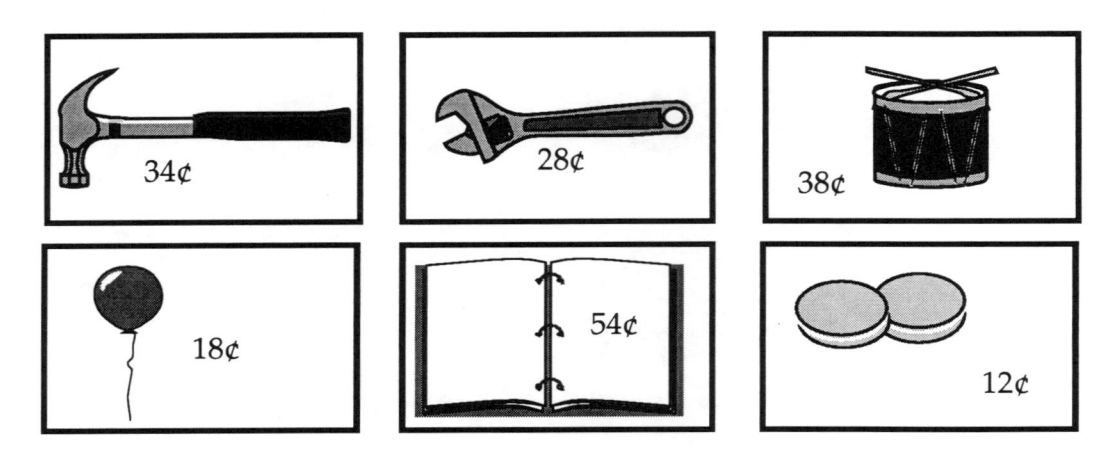

2. Two identical decks of 3 x 5 index cards with coins drawn on them, enough for each pupil (There should be two corresponding amount cards for each picture card.)

NOTE: *The exact amount of picture cards and coin cards can vary, depending on the size of the class and the amount of time available in which to play the game. At the very least, there should be one picture card for each pupil in the class.*

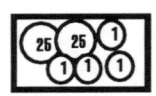

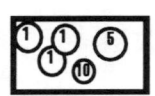

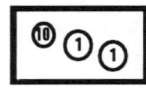

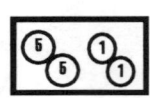

Objectives:
As a result of this activity, the learner will:
- become familiar with the names and values of the coins
- realize that one amount can be represented by different combinations of coins.

Introduction:
1. Divide the class into two teams (perhaps down the center of the classroom, so that the pupils may remain seated).

Activities for Elementary School Mathematics   Page 9.11

2. Shuffle the two decks of coin cards and distribute one deck among each team of pupils.
3. Shuffle the picture cards and place them in a stack.
4. Explain to the teams that you will show them pictures of different items and that each item will be priced. The object of the game is to search the coin cards for an amount equal to the price of the items that you will display. Make sure the teams know that one person from each team will have the correct amount on a coin card.

Major Instructional Sequence:
1. Hold up the first picture card.
2. The pupils look at their coin cards to see if they have a card with coins on it worth 28 cents.
3. If a pupil has one of the two 28-cent coin cards, he or she will stand up.

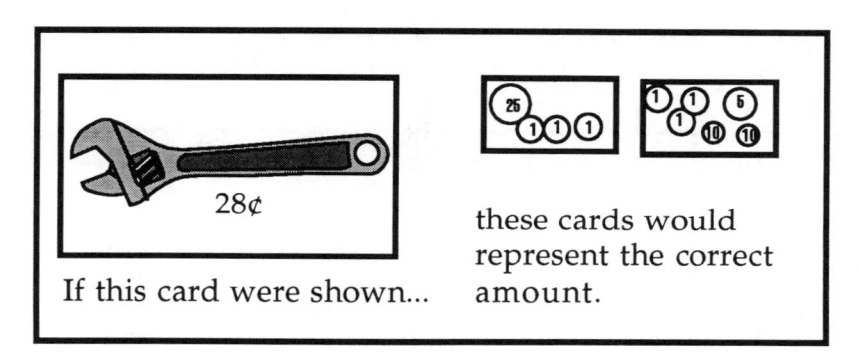

28¢
If this card were shown... these cards would represent the correct amount.

Closure or Evaluation:
Pupils score 3 points for being the first of two possible pupils to stand up and 1 point for the second response with the correct amount. The winning team is the one with the highest total score at the end of the activity.

NAMES

Topic: DIFFERENT NAMES FOR NUMBERS

Grade Level: 2-6 **Activity Time:** 1 class period

Materials Needed:
1. 3 x 5 index cards inscribed with numerals which are appropriate for the grade level (5 or 6 cards per pupil)
2. 3 x 5 cards inscribed with other names for the selected numerals

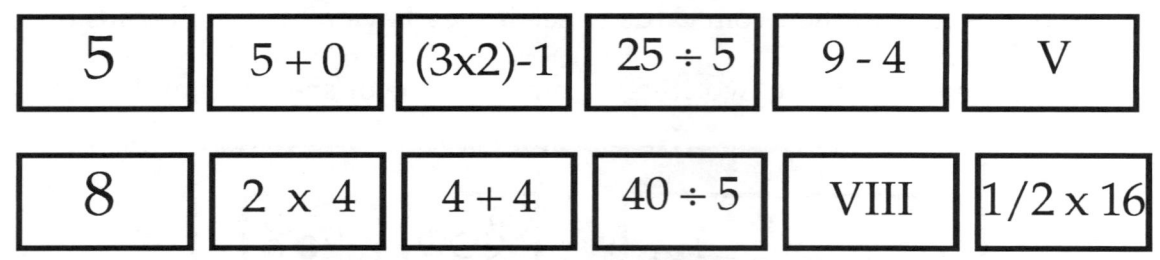

| 5 | 5 + 0 | (3x2)-1 | 25 ÷ 5 | 9 - 4 | V |

| 8 | 2 x 4 | 4 + 4 | 40 ÷ 5 | VIII | 1/2 x 16 |

3. A master list of all the correct matches

Objectives:
As a result of this activity, the learner will:
- realize that numbers have many different appearances and that there is not just one way to express each; depending on how we need to use the number, its appearance may change.

Introduction:
1. Before class, select the numbers to be reviewed for the lesson.
2. Inscribe one set of index cards with these numbers and the second set of cards with other names for the numbers.

Major Instructional Sequence:
1. Distribute 5 or 6 cards to each pupil.
2. When you hold up a number from the first set of cards (5, for example), the pupils will look through their cards for other names for 5. Those pupils who believe they have matches hold up their cards for your verification.

Closure or Evaluation:
One point is scored for each match.

Variation:
1. Pupils who have matched all their cards correctly may enter round two. (A more difficult set of cards could be created for the second round.)
2. Teams can be formed to compete for points (4 or 5 cards per round).

Activities for Elementary School Mathematics

Page 9.13

LARGE NUMBERS

Topic: READING LARGE NUMBERS

Grade Level: 4-6

Activity Time: 1 class period

Materials Needed:
1. Approximately 100 3 x 5 index cards with five- to nine-digit numbers written on them (enough for five per pupil)
2. Individual laminated "houses" and erasable markers (or sheets of "houses" reproduced on a copy machine), one for each pupil
3. A list of the numbers that corresponds with the numbers on the index cards

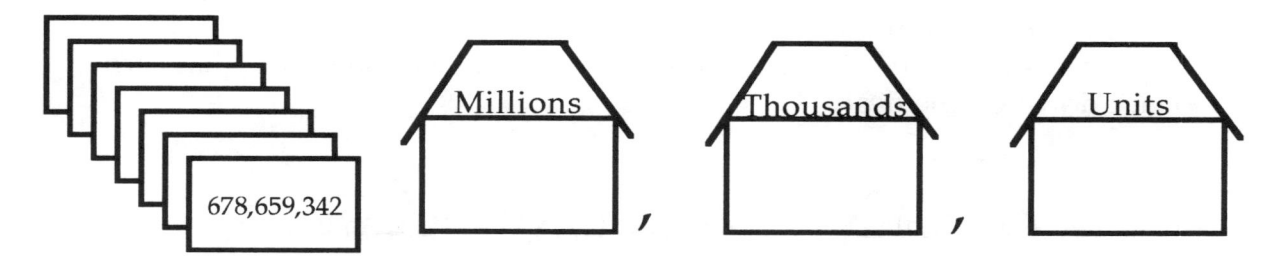

Objectives:
As a result of this activity, the learner will:
- read large numbers.
- write large numbers.

Introduction:
1. Before class, prepare a set of index cards containing five-digit to nine-digit numbers that match those on your master list.
2. Place the following drawing on the chalkboard and explain how to read it.

This number would be read as six hundred seventy-eight MILLION, six hundred fifty-nine THOUSAND, three hundred forty-two.

3. Tell the class that each house contains a family: the one to the left is the millions family; there could be a maximum of three members in the family or only one member. If the family has no members, its house would not be part of the number when it is read. Commas can be thought of as fences

between the houses to separate their yards.

Major Instructional Sequence:
1. Distribute five index cards and one marker and laminated card (or copied sheet) to each pupil taking part in the activity.
2. Instruct the pupils to lay their cards face up on their desks in front of them.
3. Select a number from the master list and read it digit by digit, as follows: 6-7-8-6-5-9-3-4-2.
4. The pupil that has the matching number raises his or her hand and attempts to read the number correctly: "I have six hundred seventy-eight million, six hundred fifty-nine thousand, three hundred forty-two."
5. If the pupil has difficulty reading the number, ask the pupil to write the numerals in the houses on his or her sheet or laminated card before trying again to read the number.
6. When the pupil successfully reads the number, he or she turns that card face down on the desk.

```
   Millions        Thousands         Units
      6      ,       420       ,      175
```

The number above would be read as
six million, four hundred twenty thousand, one hundred seventy-five.

NOTE: *The word AND does not appear in the reading of any of the numbers. AND indicates that a decimal point is in the number. 21.632 would be read as twenty-one and six hundred thirty-two thousandths.*

Closure or Evaluation:
The activity continues until all numbers have been read.

Variation:
Read the first number from the master list in the above-mentioned, digit-by-digit fashion. The pupil who holds the matching card and responds by saying the number correctly is allowed to read the next number from the list. The pupils will not know who has which numbers, so a random selection of pupils will be chosen to read their numbers.

Activities for Elementary School Mathematics

INTEGER GOLF

Topic: ADDITION AND SUBTRACTION OF INTEGERS

Grade Level: 5-6

Activity Time: 1 class period

Materials Needed:
1. A pair of number cubes for each group of 3 or 4 pupils

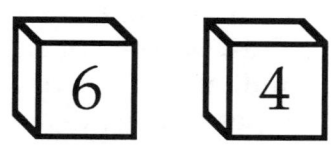

2. Two sheets of paper and a pencil for each player

Objectives:
As a result of this activity, the learner will practice:
- basic addition.
- basic subtraction.
- addition of integers.

Introduction:
1. Divide the pupils into groups of three or four. Distribute the dice to each group. Each pupil should roll one die, and the player with the highest roll begins the activity.
2. Distribute two sheets of paper and a pencil to each pupil so that everyone may do their own work and keep their own score.
3. Tell the pupils to use one sheet of paper to create individual score sheets (like the one below). Pupils will use their second sheets of paper as worksheets on which to do the actual adding and subtracting before recording their answers on their score sheets.

FRONT NINE	BACK NINE
1.	10.
2.	11.
3.	12.
4.	13.
5.	14.
6.	15.
7.	16.
8.	17.
9. ____	18. ____

TOTAL _____

Major Instructional Sequence:
1. Instruct the first pupil in each group to roll one die and record the number rolled on his or her worksheet. Tell the same pupils to roll the second die and record the number of this roll directly below the first one on their worksheets and add the two numbers for a sum. The same pupils roll again in the same way, one die at a time, and add the two numbers for a sum.
2. Direct the pupils to subtract the second sum from the first sum. The remainder is their score for the "first hole." Pupils should know that if the second sum is larger than the first one, a negative number will result.
3. Pupils are to record their scores, positive or negative, on their scorecards.
4. Play continues to the right until all players have played 9 holes of golf. (Players can play nine or eighteen holes, depending upon the time available for the mathematics activity.)

Closure or Evaluation:
1. Each pupil tallies his or her score and records it at the bottom of the scorecard for the front nine.
2. Play continues for the back nine. This score will be added to the front nine score for a total score. The player with lowest score from each group is the winner.

A sample scorecard and worksheet appear below.

FRONT NINE	BACK NINE
1.	10.
2.	11.
3.	12.
4.	13.
5.	14.
6.	15.
7.	16.
8.	17.
9. ___	18. ___

TOTAL _____

```
  6        5        3        2
  4        6        4        3
 10       11        7        5
 11                 5
 -1                 2
```

Activities for Elementary School Mathematics

SPORTS CARD STATISTICS

Topic: STATISTICS

Grade Level: 5-6

Activity Time: 1 class period

Materials Needed:
1. Paper and pencil for each pupil
2. Baseball, football, and/or basketball cards with statistical information on the back of the cards (Individual cards, team sets, pupils' favorite players or an all-star team with one player per position may be used.)
3. A set of questions which you create before class, using the statistical information obtained from the back of sports cards (one copier-reproduced question sheet per pupil). Sample cards with "STATS" appear below.

```
587              MICKEY MORANDINI                    2B
              HT. 5' 11"   WT. 170
          BORN 4-22-66 KITTANNING, PA
 YR   TEAM     AVE   G    AB    R    H   2B  3B  HR  RBI
 90   PHILLIES .241  25   79    9   19   4   0   1    3
 91   PHILLIES .249  98  325   38   81  11   0   1   20
```

```
113                  JIM KELLY                       QB
     COLLEGE: MIAMI, FLA.        HT. 6-3      WT. 219
     BORN: FEB. 14, 1960    HOME: E. BRADY, PA
                           PASSING
     YR   TEAM   ATT   COMP   PCT    YDS   TD'S   INT
     91   BILLS  346   219   63.3   2829    24     9
     92   BILLS  474   304   64.1   3844    33    17
```

Objectives:
As a result of this activity, the learner will:
- find percentages.
- find averages.
- find totals.
- figure the mean, median, mode, and range of a set of numbers.

Introduction:
1. After a unit on statistics covering mean, median, mode, and range (or other

mathematical questions) distribute the sports cards and instruct the class to find the answers to the questions you have prepared. A list of suggested questions that can be created using the "stats" appears below.

In 1990 second baseman Mickey Morandini had 19 hits in 79 at bats. What was his batting average?
(To solve this problem, the pupil must figure the batting average and then look up Morandini's average for 1990 on the back of his baseball card.)

Jim Kelly completed 304 passes out of 474 attempts in 1991. Find his completion percentage.
(To solve this problem, the pupil must divide the number of Kelly's completions by the number of his attempts in order to find the percentage completed.)

Randy Johnson gave up 61 earned runs in 1994 in 172 innings pitched. What was his earned run average?
(To solve this problem, the pupil must divide 61 earned runs by 172 innings pitched and then multiply by 9 [the number of innings in a typical game] to find Johnson's ERA.)

2. The amount and types of questions you use are limited only by your creativity and the quantity and variety of sports cards you use:
   - What is the average weight of the players on the 1994 Pirates (if a 1994 Pittsburgh Pirate team set is in the classroom)?
   - Find the difference between the highest and lowest team batting average for the 1994 Atlanta Braves (if the 1994 Braves team set is in the classroom).
   - What is the total weight of the nine starting Braves in 1991?

Major Instructional Sequence:
Distribute the questions to the pupils and permit them to work in groups of 2 or 3 to solve the questions.

Closure or Evaluation:
Pupils can check their answers by looking at appropriate cards, if the information is there, or they can share their findings with the class.

Variation:
1. This could be a two- or three-day activity, depending on the amount of pupil interest.
2. Ask groups of pupils to generate their own lists of questions. They could be shared with the class or traded between groups for discussion and solution.

Activities for Elementary School Mathematics

PROBABILITY

Topic: PROBABILITY

Grade Level: 5-6

Activity Time: 1 class period

Materials Needed:
1. A newspaper for each group of three pupils
2. Paper and pencil for each pupil in the class
3. A pair of scissors for each group

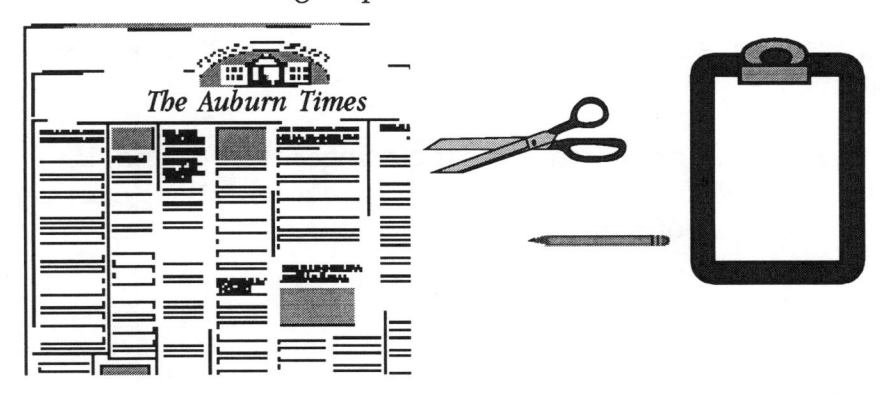

Objectives:
As a result of this activity, the learner will:
- become more familiar with probability.
- understand how to predict which letter will appear most often in a paragraph.

Introduction:
1. Tell the class that they are going to predict which two letters of the alphabet will occur most often in a paragraph in the newspaper and which two will occur the least.
2. Record their guesses on the chalkboard.

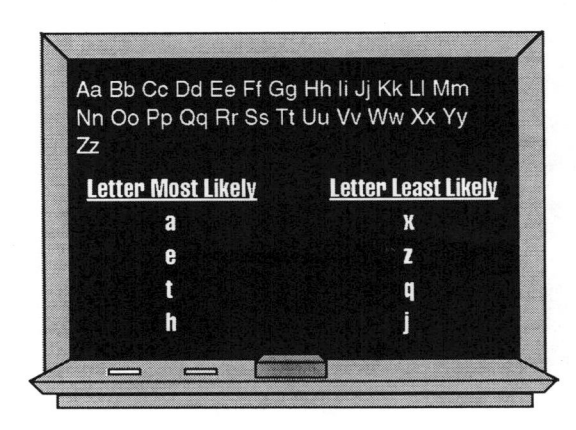

3. Distribute newspapers and ask each group to select a paragraph in which to count letters as they appear.

Major Instructional Sequence:
1. Ask each group to make a chart of the letters of the alphabet on which to record their findings.
2. Instruct the pupils to count the occurrence of individual letters (a, b, c, etc.) in their paragraphs.
3. On the chalkboard draw a master chart to record all groups' findings.

Closure or Evaluation:
Instruct the pupils to compare the actual results to their guesses.

Variation:
Pupils could guess on which letter their finger would land, if they randomly pointed to a letter in the newspaper with their eyes closed. What would be the odds of their landing on a vowel?

Activities for Elementary School Mathematics   Page 9.21

SPINNER PROBABILITY

Topic: PROBABILITY

Grade Level: 5-6   **Activity Time:** 1 class period

Materials Needed:
1. A spinner for each group of two pupils
2. A chart for each group, with a column representing each number that appears on the spinner

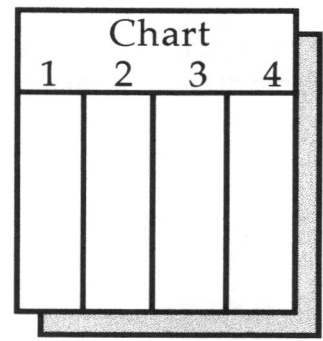

Objectives:
As a result of this activity, the learner will:
- have a better understanding of probability.
- record and analyze results of the activity.

Introduction:
1. Separate the pupils into groups of two and distribute a spinner and chart for each group. Let the pupils examine the spinners and familiarize themselves with the numbers on them. Tell the class that they will predict which of the numbers will occur most often and which will occur least often. Ask pupils to explain their reasons for their selections, based on the spinner in use.

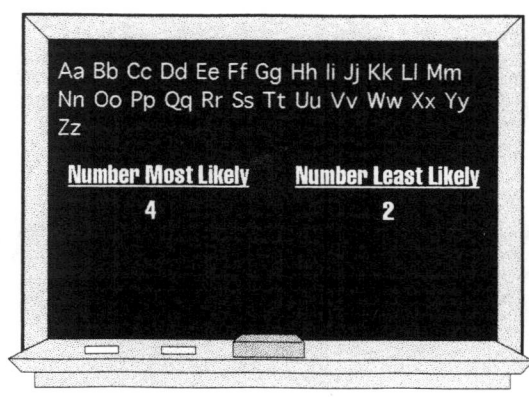

2. Instruct one pupil per pair to spin and the other to record the results. Each

pair of pupils will continue taking turns spinning until a single number comes up twenty times.

Major Instructional Sequence:
1. One pupil should spin ten times, with the other pupil recording the results on the chart.
2. After ten spins, the pupils should switch jobs and keep spinning in shifts of ten until one number occurs twenty times.
3. Ask each group to compare its results to the pupils' predictions.

Closure or Evaluation:
1. Compare the results of individual groups to the total group results.
2. On the chalkboard make a master results chart for the totals from all groups and see what happens.

Variation:
Use several different spinners with variations in the sizes allotted to each number. (See examples below.)

Activities for Elementary School Mathematics   Page 9.23

MIDTERM TEST STATISTICS

Topic: STATISTICS

Grade Level: 5-6   **Activity Time:** 1 class period

Materials Needed:
1. Paper and pencil for each pupil
2. A calculator for each group of four pupils
3. A test-results chart for each group (All charts have the same data: names of imaginary pupils, each pupil's sex, and each pupil's test score.)

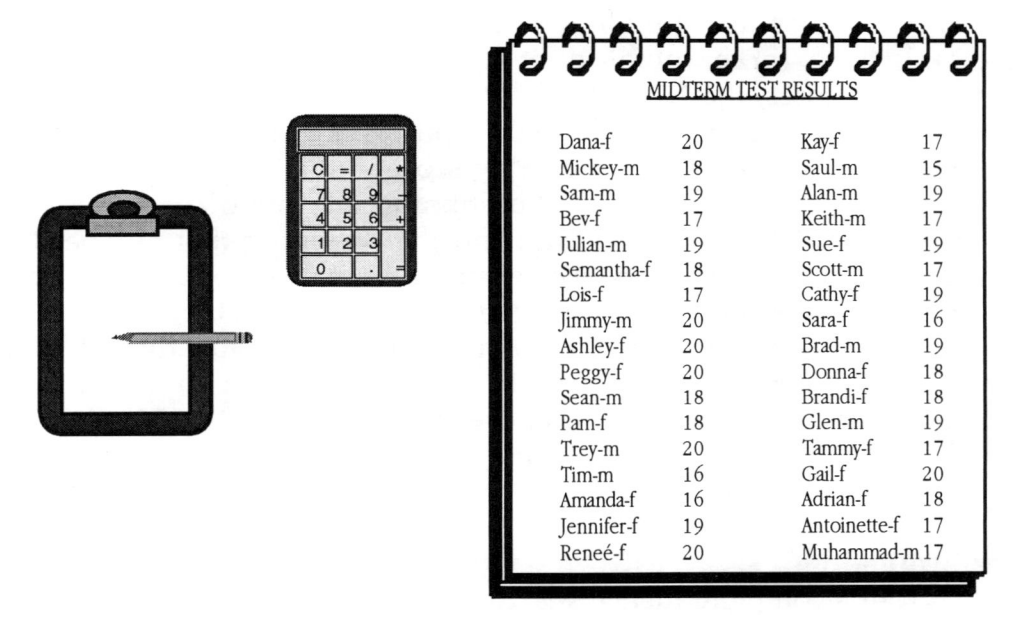

MIDTERM TEST RESULTS

Name	Score	Name	Score
Dana-f	20	Kay-f	17
Mickey-m	18	Saul-m	15
Sam-m	19	Alan-m	19
Bev-f	17	Keith-m	17
Julian-m	19	Sue-f	19
Semantha-f	18	Scott-m	17
Lois-f	17	Cathy-f	19
Jimmy-m	20	Sara-f	16
Ashley-f	20	Brad-m	19
Peggy-f	20	Donna-f	18
Sean-m	18	Brandi-f	18
Pam-f	18	Glen-m	19
Trey-m	20	Tammy-f	17
Tim-m	16	Gail-f	20
Amanda-f	16	Adrian-f	18
Jennifer-f	19	Antoinette-f	17
Reneé-f	20	Muhammad-m	17

Objectives:
As a result of this activity, the learner will:
- find percentages.
- find averages.
- find totals.
- create a frequency distribution.
- figure the mean, median, mode, and range of a frequency distribution.

Introduction:
1. After a unit on statistics covering the frequency distribution, mean, median, mode, range (or other mathematical questions), divide the pupils into groups of four.
2. Supply each group with paper and pencil, a calculator, and a test-results chart.
3. Tell the pupils they are going to work together in groups to find specific statistical data using test scores earned by pupils in a mythical class.

Page 9.24											Activities for Elementary School Mathematics

Major Instructional Sequence:
1. Use the chalkboard to demonstrate how a frequency distribution is developed from a set of test scores. (Scores are listed in the frequencies that they occur, starting with the highest score. See example below:)

   20
   20
   20
   19
   19
   18
   18
   etc.

2. Use the frequency distribution to show pupils how to find the median. (The median is the exact midpoint in the frequency distribution. For example, if there are 25 scores, the 13th score is the median.)
3. Use the frequency distribution to show pupils how to find the mode. (The mode is the most frequently occurring score in the frequency distribution.)
4. Tell pupils that the mean is found by finding the sum of all of the scores in the frequency distribution (adding) and then dividing that number (the sum) by the total number of scores. For example, if the sum of 25 scores in the frequency distribution is 425, then 425 ÷ 25 = the mean.)
5. Using their test-results charts, instruct pupils to work together collaboratively to create a frequency distribution, then find the median, mode, and mean. (The calculators may be used whenever necessary.)
6. Circulate among the pupils, checking for understanding, and providing assistance when appropriate.

Closure or Evaluation:
1. As groups complete their tasks, direct them to find three additional pieces of information using their data (The calculators may be used whenever necessary.):
   - Which group had the highest mean score, males or females?
   - What percentage of females made a perfect score among females; among the entire class?
   - What percentage of males made a perfect score among males; among the entire class?
2. Lead the groups in a discussion where they share their findings and discuss some of the difficulties they encountered in making their computations.

CHAPTER TEN
PROBLEM SOLVING

According to the National Council of Teachers of Mathematics, one of the main objectives of a mathematics teacher (or *any* teacher) should be to make pupils problem solvers. Pupils should be more than just mathematics problem solvers, they should become solvers of all sorts of problems in their daily lives. Pupils will be confronted with problems daily and need to be taught strategies that will help them solve these problems.

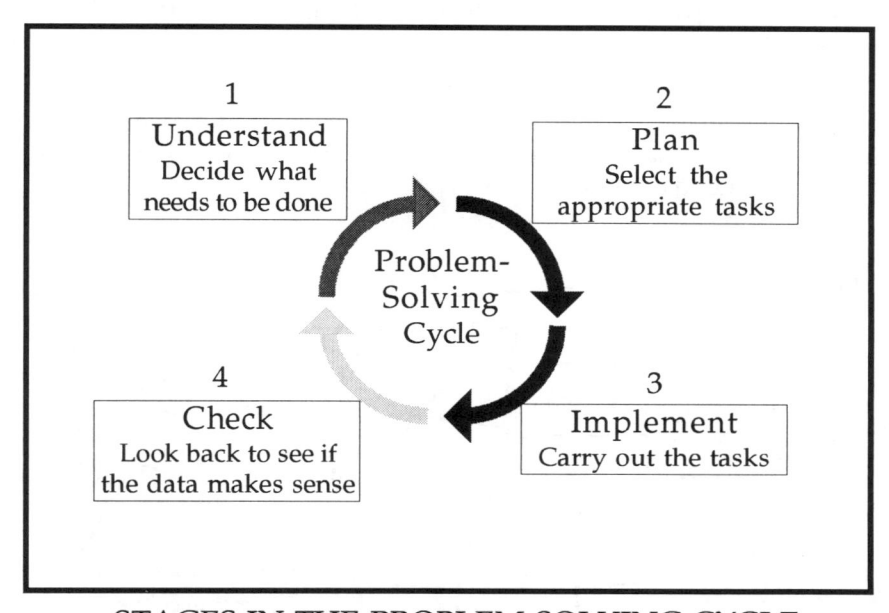

STAGES IN THE PROBLEM-SOLVING CYCLE

1. **Understand** — Decide what needs to be done
2. **Plan** — Select the appropriate tasks
3. **Implement** — Carry out the tasks
4. **Check** — Look back to see if the data makes sense

Whenever problem solving is mentioned among mathematics teachers, George Polya[1] and his book *How to Solve It* (1957) usually comes to mind. His strategy can be boiled down to four basic steps which should help pupils become problem solvers.

The first step is to *realize that there is a problem that must be solved.* You leave the store late at night and your car is the only one in the parking lot, and it will not start. You realize you have a problem! What are you going to do?

Step two is to *devise a plan to solve your problem.* After you stop crying, you decide that you are going to call your husband, your wife, your father, or a taxi and get a ride home.

The third step is to *put your plan into action.* It does no good to come up with a plan and stay in your car; you must put the plan into action and call whomever you felt would be able to help get you out of the mess you are in at this time. If the plan does not work, you must revise it or come up with a new plan.

---
[1]Polya, G. (1973). *How to solve it.* 2nd ed. Princeton, N.J.: Princeton University Press.

Page 10.2   Activities for Elementary School Mathematics

Finally, the last step is to *make sure the plan worked*. George Polya called this "looking back." Did my plan get the results that I desired?

In mathematics, this involves more than checking individual calculations to ensure that they were accurately performed. It should also include making sure the answer makes sense — and, if it does not, checking whether the correct calculations were chosen to solve the problem. It is possible to perform all of the mathematical operations correctly, yet the answer will not be correct. An example of this follows:

A bus was going from Phoenix, Arizona to Chicago, Illinois, a distance of approximately 1,750 miles. The bus driver wanted to make the trip in 5 days and spend about 7 hours a day traveling. What speed must the driver average to accomplish this goal?

Some pupils multiplied 1,750 x 5 x 7 and got an answer of 61,250 miles per hour. The pupils realized that the driver on the hypothetical bus trip would be travelling a bit too fast! They checked their multiplication and found they had multiplied correctly, but the answer did not make sense. The correct answer includes two division steps, not just multiplication. Pupils need encouragement, not only to check that the calculations are done correctly, but also to make sure that the correct calculations are chosen so that the answer makes sense.

Activities for Elementary School Mathematics   Page 10.3

SCORE 33

Topic: PROBLEM SOLVING

Grade Level: 4-6   **Activity Time:** 1 class period

Materials Needed:
1. Chalkboard and chalk
2. Paper and pencil for each pupil

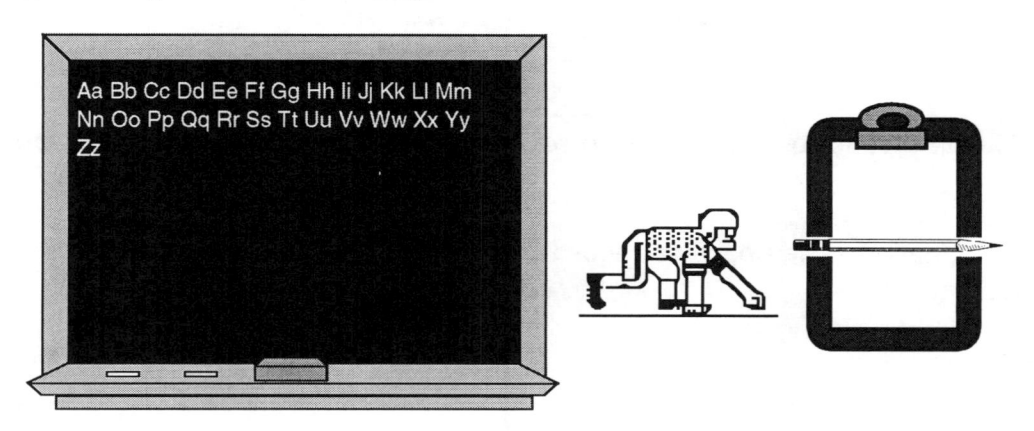

Objectives:
As a result of this activity, the learner will:
- be a better problem solver.

Introduction:
1. Tell the class that a professional football team scored 33 points in their last game.
2. Tell the pupils that they must determine as many ways as possible for the team to have scored their 33 points.

Major Instructional Sequence:
1. Pupils should first decide the ways in which a professional football team could score points. Discuss the following rules for scoring and write them on the chalkboard for pupils to use as a reference:
   - 6 points for a touchdown
   - 7 points for TD and extra point
   - 3 points for a field goal
   - 8 points for TD & *2 point conversion
   - 2 points for a **safety

   *A team receives 2 points after a TD if they run or pass and score.
   **A safety occurs when a team is downed behind their own goal line.

Page 10.4   Activities for Elementary School Mathematics

> Aa Bb Cc Dd Ee Ff Gg Hh Ii Jj Kk Ll Mm
> Nn Oo Pp Qq Rr Ss Tt Uu Vv Ww Xx Yy
> Zz
> - 6 points for a touchdown
> - 7 points for TD and extra point
> - 3 points for a field goal
> - 8 points for TD & *2 point conversion
> - 2 points for a **safety
>
> *A team receives 2 points after a TD if they run or pass and score.
> **A safety occurs when a team is downed behind their own goal line.

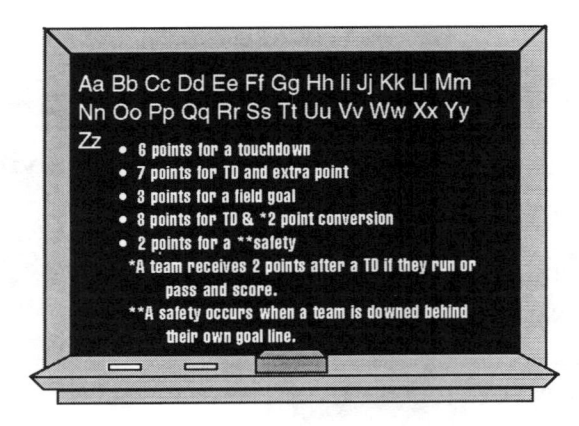

2. Ask the pupils to come up with several ways a team could score 34 points.

Closure or Evaluation:
1. Put a chart on the board to record pupil responses and to check their calculations.

	2 pts.	3 pts.	6 pts.	7 pts.	8 pts.
Example 1	I	I	I	I	I I
Example 2			I	I I I I	

2. In example one, pupils solved the problem by 2 (points for a safety) + 3 (points for a field goal) + 6 (points for a touchdown) + 7 (points for a TD and extra point) + 8 (points for a TD and two-point conversion) + 8 (again) = 34. In example two, pupils solved the problem by 6 (points for a touchdown) + 4 x 7 (points for TDs and extra points) = 34 points.

Activities for Elementary School Mathematics   Page 10.5

ZOO

Topic: PROBLEM SOLVING

Grade Level: 4-6   **Activity Time:** 1 class period

Materials Needed:
   Paper and pencil for each pupil

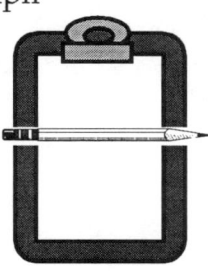

Objectives:
   As a result of this activity, the learner will:
   - be a better problem solver.
   - function as part of a group to find solutions.

Introduction:
   1. Relate the following problem to the class and write it on the chalkboard: When two pupils went to a zoo, one pupil saw some bears and some pelicans. The other pupil saw a total of 52 legs when they were there. How many of each animal did the pupils see at the zoo?

   2. Ask the class to determine how many of each *may* have been seen. (Bears are four-legged creatures.)

Major Instructional Sequence:
   1. Divide pupils into groups of three or four. Instruct the groups to find as many possible answers as they can.
   2. Permit the pupils to work until each group has three or four answers. Groups may make a chart to solve the problem, if necessary.

Page 10.6   Activities for Elementary School Mathematics

Closure or Evaluation:
1. Discuss the possible answers with the class, with each group given an opportunity to give at least one answer.

Bears	Pelicans
1	24
2	22
3	20
...	...
11	4
12	2

2. Remember that some of each animal were seen, so zero is not a possible answer for either group.

Activities for Elementary School Mathematics

PRICE CHECK

Topic: PROBLEM SOLVING

Grade Level: 4-6

Activity Time: 1 class period

Materials Needed:
1. Weekly grocery-store advertisements from three or four local stores
2. Paper and pencil for each group of pupils
3. A calculator for each group

Objectives:
As a result of this activity, the learner will:
- solve problems based on prices and decide, through comparative shopping, in which store to shop for groceries.

Introduction:
1. Before class, write on the chalkboard a grocery list of fifteen to twenty items which can be found on several (preferably most) of the stores' advertisements.

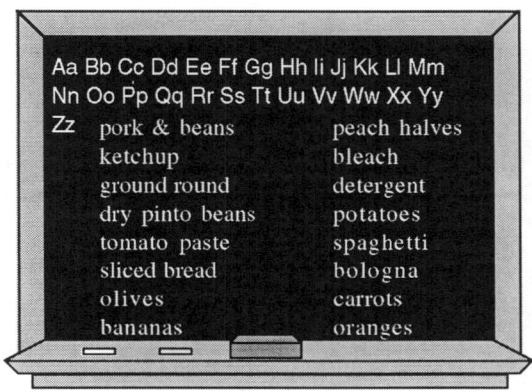

NOTE: *Several factors to consider when selecting items for pupils to find and price are different weights and sizes, different quantities, different brands, etc.*

2. Separate pupils into five groups.
3. Instruct the groups to copy the list and obtain prices from the advertisements. Tell pupils that they are comparison shopping and therefore should obtain as many different prices as possible for the items on their shopping list.

Item	Store A	Store B	Store C
pork & beans	3/$1.00	$.39	2/$.75
peach halves	$.98	2/$2.00	$1.12
ketchup	28 oz. $.88	32 oz. $.99	28 oz. $.87
bleach	$.88/gal	$.69/gal	$.79/gal
ground round	$1.67/lb.	$1.60/lb	3lbs/$4.50

Major Instructional Sequence:
1. Instruct the groups to trade advertisements with other groups so they can find the items and their prices from the new advertisements.
2. Direct the groups to keep trading the ads until they have had access to all the store advertisements.
3. Ask the groups to decide which store offers the best price on each item and how much they would save buying there.
4. The groups should decide which store has the best overall prices.

Closure or Evaluation:
Discuss with the class how such factors as weight, size, quantity, and brand may affect the price of groceries. Discuss whether going to each store for different items is a possibility, and discuss the pros and cons of doing that.

Activities for Elementary School Mathematics

Page 10.9

BALANCED ANSWER

Topic: PROBLEM SOLVING

Grade Level: 4-6

Activity Time: 1 class period

Materials Needed:
1. Eight boxes of identical size, with seven boxes weighing the same (The weight of one box should be more than that of each of the others.)

2. A balance scale large enough to weigh the boxes
3. Eight 3 x 5 index cards
4. Paper and pencil for each pupil

Objectives:
As a result of this activity, the learner will:
- problem solve.
- become familiar with a balance scale.

Introduction:
1. Show the class the eight boxes and tell them one of the packages is heavier than the others.
2. Explain that the object of the exercise is to find which is the heavier package, using the balance scale only three times.

Major Instructional Sequence:
1. Ask pupils to work in groups and discuss how they could solve the problem.

Page 10.10   Activities for Elementary School Mathematics

2. Permit the groups time to discuss and record on paper their attempts to solve the problem.
3. The first group that comes up with a solution should demonstrate the solution to you (to verify their correct response). If the solution is correct, do not tell the class but permit another group to test the first group's solution.

Closure or Evaluation:
The actual solution is illustrated below:
1. Put four packages on each side of the scale and see which side weighs the most. The heaviest package will be on that side of the scale. Put the four lighter packages aside.

2. Divide the four packages from the heavier side into two groups of two, and put one group on either side of the scale.

3. Two of the packages will be heavier than the other two. Set the lighter two packages aside and put the remaining two on the scale. The heaviest of the packages will be obvious at this time.

Activities for Elementary School Mathematics

BARGES OR TRUCKS?

Topic: PROBLEM SOLVING

Grade Level: 4-6

Activity Time: 1 class period

Materials Needed:
1. Paper and pencil for each pupil
2. One typewritten problem sheet for each group of four

Objectives:
As a result of this activity, the learner will:
- work in a collaborative group to solve problems.
- chart the steps necessary for finding a solution.

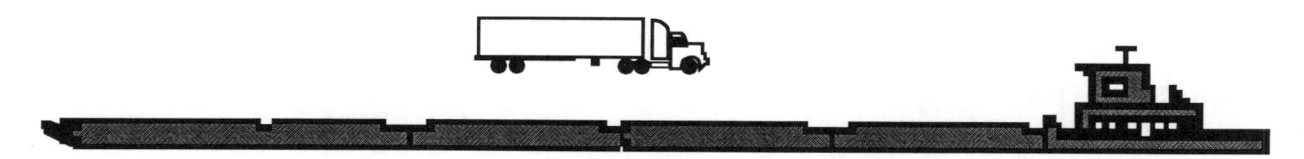

In preparation for this activity, place the following information on the chalkboard:
- One barge = 60 tractor-trailor loads
- Baton Rouge to Memphis is 381 miles by highway but only 300 miles via the Mississippi River.
- Tractor-trailor trucks average 8 miles per gallon of diesel fuel with a maximum load.
- Towboats average 3 miles per gallon of diesel fuel with a maximum load (15 barges) when headed upstream.

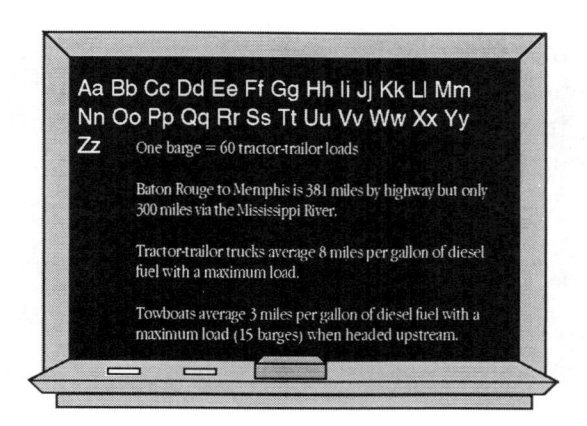

Introduction:
1. Divide the class into groups of four.
2. Give each group the following problem situation typewritten on a sheet of paper:

A Baton Rouge, Louisiana, planter wants to find the most economical way to make grain shipments to Memphis, Tennessee. Your group operates towboats and barges on the Mississippi River and wants to get the planter's business. Calculate the following information for your presentation to the planter:
1. The number of tractor-trailor loads you can carry in one trip.
2. The cost of transporting that amount via tractor-trailer truck.
3. The cost of transporting that amount via a towboat and barges.
4. The savings for the planter if he ships by barge.

Major Instructional Sequence:
1. Read aloud and discuss the information on the chalkboard. (Pupils may volunteer to read aloud.)
2. Read aloud and discuss the information on the problem sheet. (Again, pupils may volunteer to read aloud.)
3. Instruct the groups to work collaboratively to solve the problems presented on the problem sheet and to plan a step-by-step presentation to be made to the planter.
4. Circulate among the groups, check for understanding, and provide assistance as needed.

Closure or Evaluation:
1. Ask each group, in turn, to give their answers to the specific questions asked on the problem sheet. Discuss and clarify, as appropriate.
2. Ask each group, in turn, to give their step-by-step presentation planned for the planter. Discuss and clarify, as appropriate.

Activities for Elementary School Mathematics — Page 10.13

SHOPPERS

Topic: PROBLEM SOLVING

Grade Level: 4-6

Activity Time: 1 class period

Materials Needed:
1. Magazines, newspapers, advertising brochures, catalogs, and the like, with a variety of items advertised in them for each of six groups
2. Scissors, glue, and poster board (about 24 x 36) for each of six groups
3. A calculator for each of six groups

Objectives:
As a result of this activity, the learner will:
- problem solve and get experience filling out an order form.
- gain experience figuring sales tax (percentages).

Introduction:
1. Separate the class into six groups.
2. Tell the class that they are opening a variety store.
3. Ask each group to clip up to 30 items (advertisements from the magazines, etc.) to "stock" the store. Tell them to be sure to include the prices of the items.
4. Have groups cluster the items in categories of a similar nature (tell them they are creating "departments" in the store).

5. Ask pupils to place similar items close to each other as they glue the ads (including prices) to the poster board, mounting them in groups.

Major Instructional Sequence:
1. Direct each group to work collaboratively and prepare a shopping list of fifteen items derived from their poster-board department store.
2. Tell the groups to compute the cost of the items in their shopping list on a separate sheet of paper, adding an 8% sales tax.
3. Have each group exchange its poster-board department store and shopping list with a different group.
4. Ask pupils to work collaboratively to "shop" for the items on their shopping list, compute the cost of the items bought, and add an 8% sales tax.

Closure or Evaluation:
1. Each group reports on its shopping "spree," telling what items were bought and how much they spent, including sales tax.
2. Their totals are verified (or disputed) by the group who made the poster board and shopping list used by the group giving the report.
3. Discrepancies are examined and resolved by the whole class in a discussion led by the teacher, showing the computations on the chalkboard or overhead.

Variation:
1. Other ideas could be incorporated into the shopping as follows:
    - Have the groups see how close they can make their list to $200 and give them a time limit.
    - Have groups plan a camping trip and buy the equipment that is necessary using catalogs for camping equipment.
    - Plan a 20%-50% off sale on items and have pupils decide which items to mark down for the sale.
2. As a separate lesson, give pupils experience in filling out an order form, such as the one in the appendix. The reproducible form in the appendix allows for the ordering of several items, tax computations with variations for different states, and shipping costs which are variable depending upon the number of items ordered.

Activities for Elementary School Mathematics

STOOL MAKER

Topic: PROBLEM SOLVING

Grade Level: 5-6

Activity Time: 1 class period

Materials Needed:
1. The following problem, typewritten and reproduced for each group

> A stool maker makes three- and four-legged wooden stools. He had 96 stool legs. He makes a profit of $4 on the three-legged stools and a $5 profit on the four-legged stools. After he used *all* 96 legs, his profit was $122. How many of each type of stool did he make?

2. Paper, pencils, and chart paper for each group of three pupils

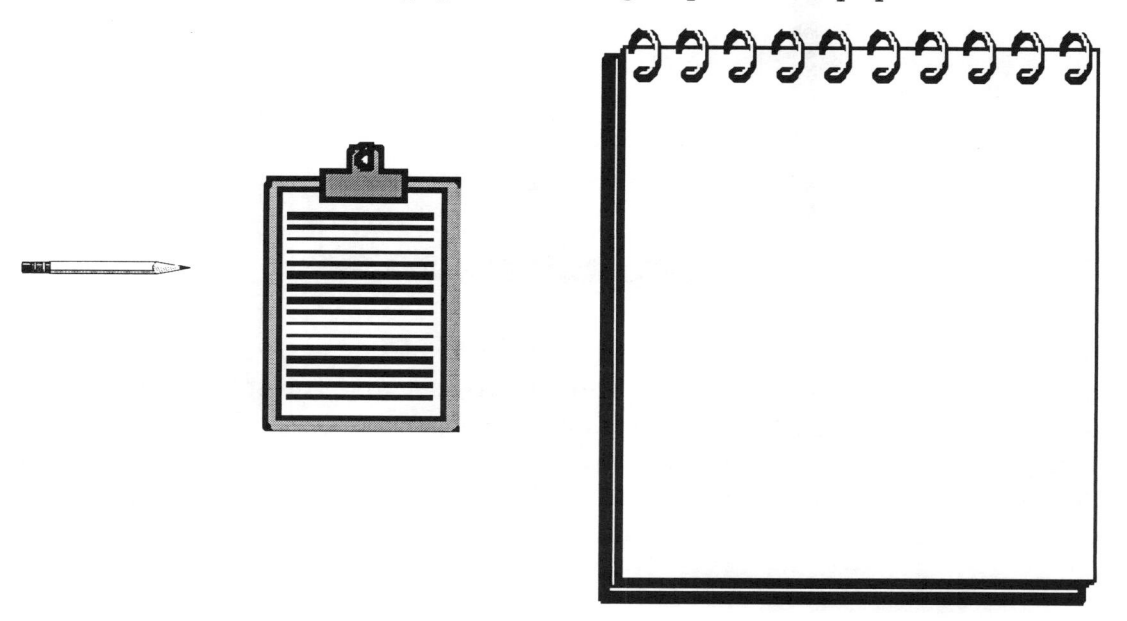

Objectives:
As a result of this activity, the learner will:
- become a better problem solver.
- make a chart to help solve problems.

Introduction:
Read the story-problem to the class and answer any questions they may have about it.

Page 10.16   Activities for Elementary School Mathematics

Major Instructional Sequence:
1. Divide pupils into groups of three. Distribute a copy of the problem to each group. Tell the pupils they should attempt to set up a chart that will help them solve the problem.
2. Circulate among the pupils, checking for understanding and giving assistance as necessary.
3. For those groups that are having difficulty in setting up a chart, suggest that the following chart be used.

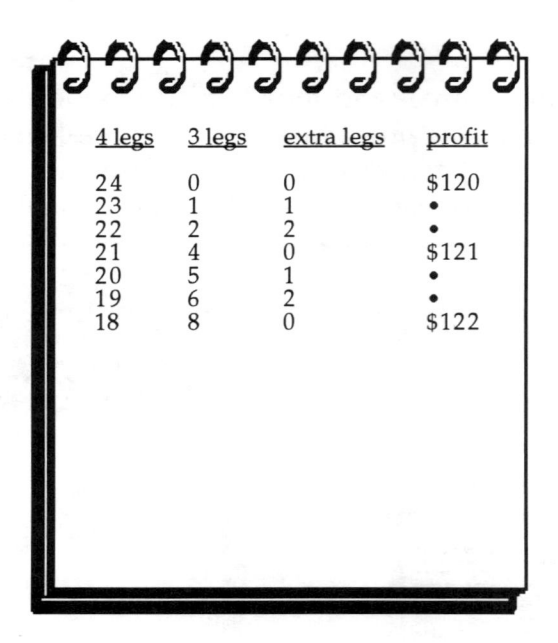

4 legs	3 legs	extra legs	profit
24	0	0	$120
23	1	1	•
22	2	2	•
21	4	0	$121
20	5	1	•
19	6	2	•
18	8	0	$122

Closure or Evaluation:
Answers with extra legs are not considered, because the stool maker used ALL the legs. One solution to the problem is: 18 (four-legged stools) x $5 =$90, and 8 (three-legged stools) x $4 =$32, for a total of $122. Another solution is: 6 (four-legged stools) x $5 = $30, and 23 (three-legged stools) x $4 = $92, for a total of $122.

Activities for Elementary School Mathematics

TOTAL 9

Topic: PROBLEM SOLVING

Grade Level: 5-6

Activity Time: 1 class period

Materials Needed:
1. One notebook sheet of paper per pupil
2. One larger sheet of paper

Objectives:
As a result of this activity, the learner will:
- improve problem-solving skills.

Introduction:
1. Hold up the large sheet of paper and instruct the class to follow your directions, using their notebook paper. Fold paper long ways down the middle and then tear down the fold, verbally reinforcing your actions so that the class can follow your example.
2. Place the two halves on top of each other, fold together into thirds (make an "S" and press down), and tear along these folds. Each pupil should have 6 pieces of paper. Tell the pupils that they will be solving a mathematical puzzle using these pieces.

Major Instructional Sequence:
1. Instruct the pupils to write a 1 on the first piece of paper, a 2 on the second, and so on until all six pieces have one numeral on them.
2. Tell them to construct a triangle (or pyramid) out of the 6 pieces of paper as follows:

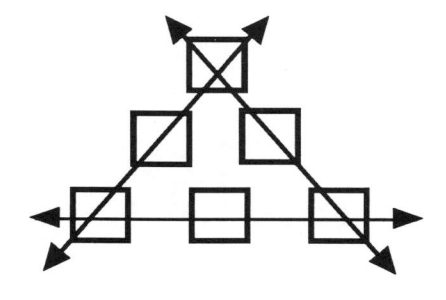

3. Their task is to arrange the pieces of paper so any three pieces in a row can be added to get a total of 9.

Page 10.18   Activities for Elementary School Mathematics

Closure or Evaluation:
   After a specified time period, ask for volunteers to present their solutions (they may write their number arrangements on the chalk board). Tell the class that there are six possible solutions. After six volunteers have drawn their solutions (or, if no volunteers are forthcoming, after you have discussed the solutions with the class), ask the pupils to look for patterns within the triangles/pyramids. For example, the first card has 1, 5, 3 ascending on the left and the third card has 1, 5, 3 descending on the right; the second card has 3, 4, 2 ascending on the left and the fifth card has 3, 4, 2 ascending on the right.

Six Possible Solutions

| 3 | 2 | 1 |
|5   4 | 4   6 | 6   5 |
|1   6   2| 3   5   1| 2   4   3|

| 3 | 2 | 1 |
|4   5 | 6   4 | 5   6 |
|2   6   1| 1   5   3| 3   4   2|

Activities for Elementary School Mathematics

Page 10.19

COLORADO MOUNTAINS

Topic: PROBLEM SOLVING AND ROUNDING

Grade Level: 5-6

Activity Time: 1 class period

Materials Needed:
1. A map of the United States

2. Colorado road maps
3. Paper and pencil for each pupil

Objectives:
As a result of this activity, the learner will:
- locate selected areas on a map
- be able to round to the nearest ten, hundred and thousand

Introduction:
1. On the chalkboard list some of the highest elevations in Colorado, for example:

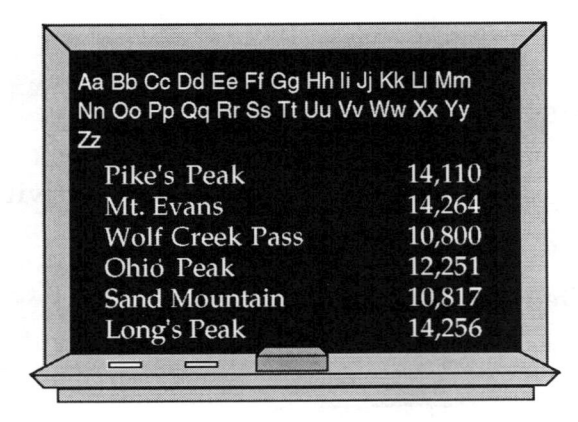

Page 10.20   Activities for Elementary School Mathematics

```
Aa Bb Cc Dd Ee Ff Gg Hh Ii Jj Kk Ll Mm
Nn Oo Pp Qq Rr Ss Tt Uu Vv Ww Xx Yy
Zz
    Pike's Peak        14,110
    Mt. Evans          14,264
    Wolf Creek Pass    10,800
    Ohio Peak          12,251
    Sand Mountain      10,817
    Long's Peak        14,256
```

  2. Tell the class that they are planning an imaginary vacation and that they will need to read maps and make some estimations in order to make the trip.

Major Instructional Sequence:
1. Separate the pupils into groups of four. Ask the groups to copy the list, beginning with the highest and ending with the lowest elevation. Distribute one road map of Colorado per group, and instruct the pupils to find each location on the maps.

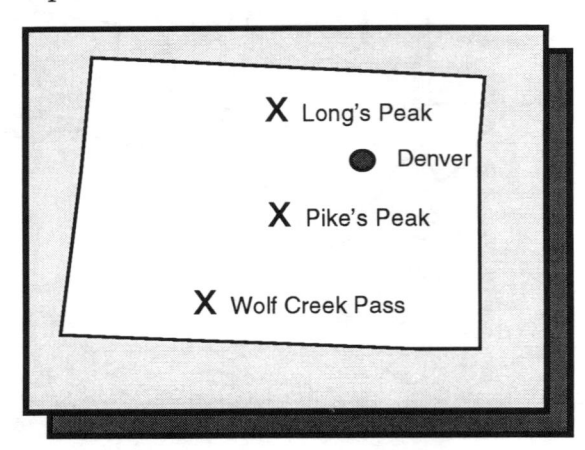

```
    X Long's Peak
        ● Denver
    X Pike's Peak

X Wolf Creek Pass
```

2. Then ask the pupils to round each elevation to the nearest ten, hundred and thousand.
3. Next, ask the class to calculate how many yards, miles, meters, etc. each elevation would be if converted to the new unit. (One unit of measure could be assigned to each group.)

Pike's Peak	
feet	14,110
yards	4703.3
inches	169,320
meters	4275.76

Activities for Elementary School Mathematics

Closure or Evaluation:
    The class shares and discusses their calculations.

Variation:
1. The class could plan a trip from your school to one of the locations by:
   - figuring the mileage from school to the location
   - estimating the number of days it would take to reach the location
2. Pupils could figure hotel and meal prices for the trip. If you can obtain copies of American Auto Association (AAA) guidebooks for class use, this imaginary vacation planning could take on new dimensions of complexity. If you provide an imaginary price per gallon, pupils can also calculate the cost of gas.

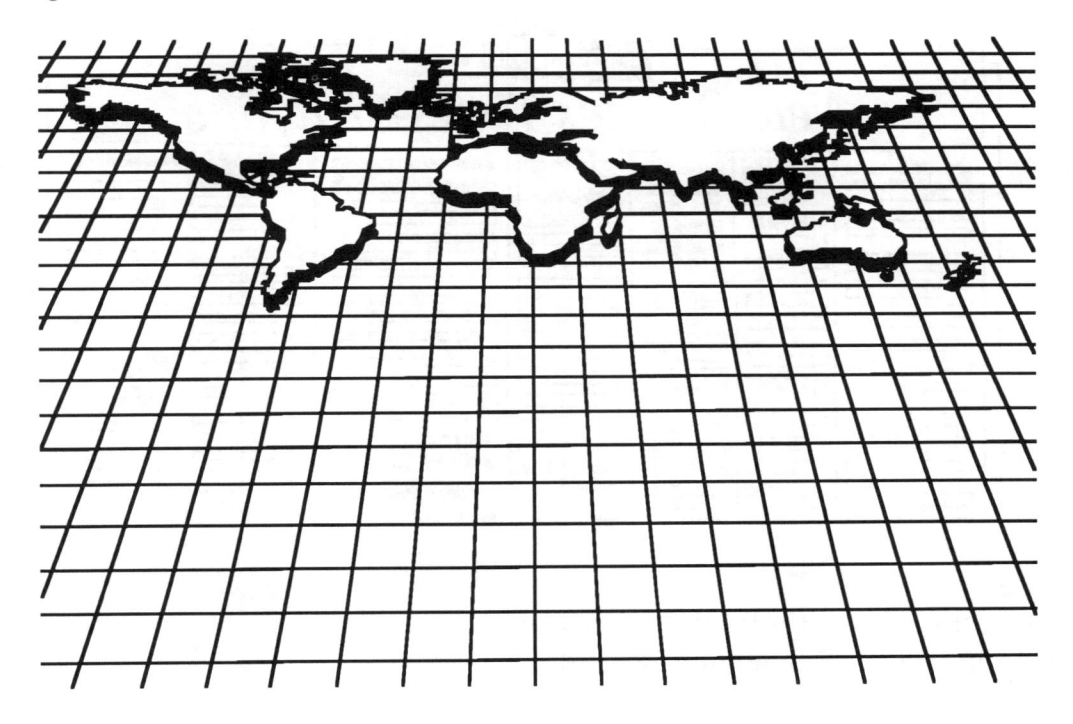

TODAY'S NEWS

Topic: PROBLEM SOLVING

Grade Level: 4-6   **Activity Time:** 1 or 2 class periods

Materials Needed:
1. Eight to ten copies of a current newspaper
2. Paper and pencil for each pupil

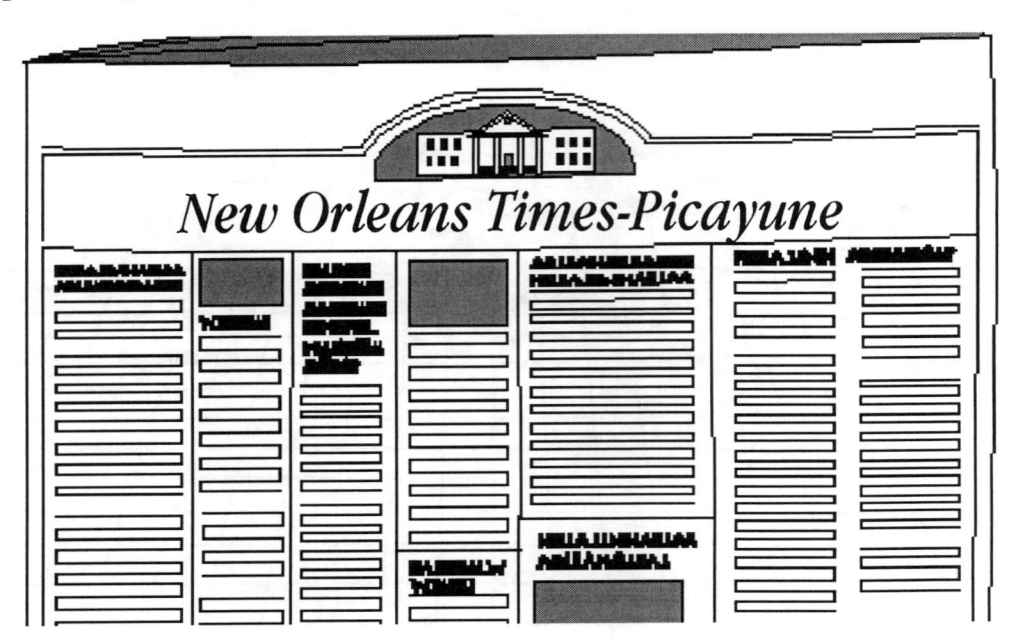

Objectives:
As a result of this activity, the learner will:
- improve problem-solving skills.
- learn to communicate mathematically.
- use critical thinking skills in evaluating story problems.

Introduction:
1. Before class time, read the paper and locate at least ten interesting facts from the daily news.
2. After presenting several of these facts to the pupils, tell them they will write story problems based on the articles which contain the facts. Tell the class not to worry about spelling or punctuation in their original drafts of the stories. Pupils will revise their stories and correct any mistakes for their final drafts.

Activities for Elementary School Mathematics   Page 10.23

Examples from the daily paper:

- The National Collegiate Athletic Association, the governing body for college football, allows Auburn University to award 26 new football scholarships each year.

    Pupil's word problems could be: How many football scholarships would a student see awarded if he stayed at Auburn for four years? OR, How many football scholarships were awarded at Auburn University between 1990 and 1997?

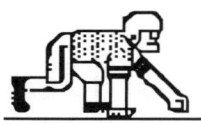

- The governor used the state plane for personal business. Each flight cost the taxpayers $1,200.

    Pupil's word problems could be: If the governor took 13 trips at taxpayers' expense, how much was the total cost? OR, If each trip was 356 miles long, how much per mile did each trip cost?

- Hurricane Luis is approximately 750 miles southeast of the United States and is traveling 12 miles per hour with winds at 140 miles per hour. Luis is traveling North Northwest.

    Pupil's word problems could be: If Luis continues at his present speed, how long would it be before the storm reaches the United States?

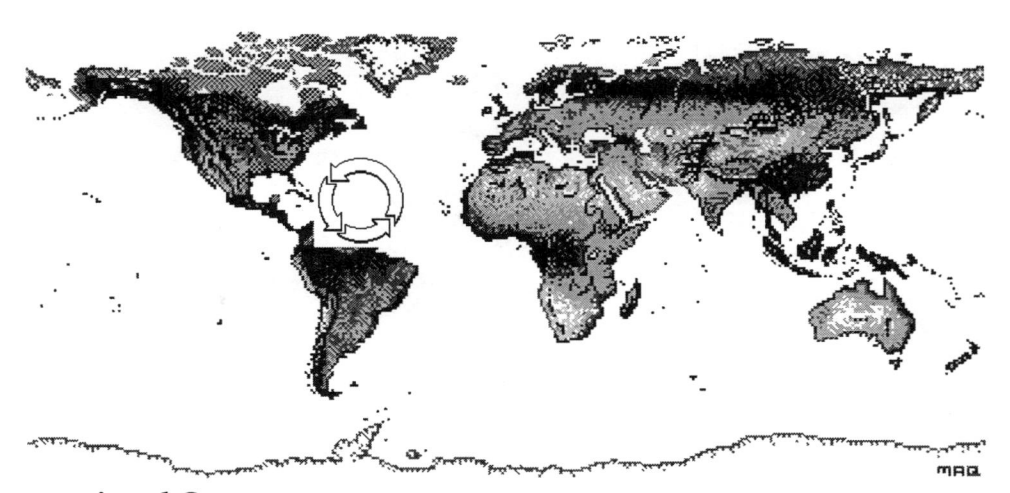

Major Instructional Sequence:
1. Place the pupils in groups of 3 or 4 and give each group a different fact/article from which to create a story problem. Each group is then to solve the problem they create.
2. Each group should take a turn to present their article/fact and their problem to the rest of the class for a solution.

Closure or Evaluation:
1. After each group makes its presentation and its problem has been solved, encourage class discussion about the problem, asking them to find some assumptions which could alter the answers, for example:

   - In the Auburn University article, suppose that Auburn was placed on probation for 1994-1995 and could offer only 15 scholarships.

   - Not all of the governor's flights were to the same airport.

   - Luis may change directions. The speed of the hurricane may change from hour to hour. Luis may not reach the United States, but just drift harmlessly out to sea.

2. As part of an English assignment, ask each group to correct the spelling and punctuation in their story problems.

Annotated Bibliography of Children's Books for Teaching Mathematics

Number Sense and Numeration—Books Suitable for Grades K-3

Emberley, Barbara. *One Wide River to Cross.* Englewood Cliffs, NJ: Prentice Hall, 1976.
> A rhyming book where animals come, one by one, two by two, three by three, etc., all the way to ten by ten to the ark built by Noah. Then it rains and rains and rains. Finally, after the rains halt, the ark winds up on Mount Ararat.

Giganti, Paul. *How Many Snails?* New York, Greenwillow, 1988.
> The author walks to various places (lake, beach, park, library, bakery, and the like) and wonders about the different characteristics of the things he sees in each place.

Krahn, Fernando and Maria de la Luz Krahn. *The Life of Numbers.* New York: Simon & Schuster, 1970.
> Number One is very bored living alone, so he looks for a playmate. First he meets Zero, but Zero is nothing and One doesn't want to play with him. One meets up with the rest of the numerals, Two through Nine, but for a variety of reasons, none of them play with One. Going back home, One meets up with Zero again, and they get together to become Ten.

Lottridge, Celia. *One Watermelon Seed.* Toronto: Oxford University Press, 1986.
> A counting book, giving children opportunities to count not only from 1 to 10, but also from 10 to 100, as they accompany Max and Josephine through the planting, weeding, watering, and harvesting of their garden.

Mack, Stan. *10 Bears in My Bed.* New York: Pantheon Books, 1974.
> A delightful goodnight countdown book. Going to bed, a little boy finds ten bears in his bed. One by one, the bears depart through the window as the little boy tells them to roll over. When all ten bears leave, the little boy has the bed to himself.

Maestro, Betsy and Giulio Maestro. *Harriet Goes to the Circus.* New York: Crown, 1989.
> A good book for introducing ordinal numbers. Harriet is an elephant who gets up early to go to the circus so that she can be first in line and get a good seat. Harriet gets there first and all of her animal friends line up one-by-one behind her. But there is a problem! The entrance to the circus tent is at the opposite end of the line. Everyone turns around, and to her dismay, Harriet is now last in line. As it turns out, however, the chairs inside the tent are in a circle and everyone gets a good seat.

Zaslavsky, Claudia. *Zero. Is It Something? Is It Nothing?* New York: Franklin Watts, 1987.
In discussing zero, the author looks at the meaning and mathematical possibilities of zero through various ways, including riddles about zero and the history of zero.

Number Sense and Numeration—Books Suitable for Grades 4-6

Adler, David. *Roman Numerals.* New York: Thomas Y. Crowell, 1977.
An introduction to the development and use of Roman numerals, including an activity to help pupils translate between Roman numerals and Hindu-Arabic numerals.

Carlson, Nancy. *Harriet's Halloween Candy.* Minneapolis, MN: Carolrhoda Books, 1982.
After getting a big bag of candy trick-or-treating, Harriet dumps the candy onto the floor and organizes it by color, size, and her favorite kinds of candies. Each time Harriet eats some of the candy, she hides the rest in different places, but when she runs out of hiding places, she decides to eat the rest of the candy.

Froman, Robert. *Less Than Nothing Is Really Something.* New York: Thomas Y. Crowell, 1973.
This book sketches assorted everyday applications of negative numbers, and includes a game for pupils to play.

Razzel, Arthur and K. G. Watts. *This Is 4: The Idea of Number.* Garden City, NY: Doubleday, 1967.
This book explores the idea of four behind the numeral 4. Dealing with sets of 4 and various names for different sets of 4, the explorations delve into ideas like quadrilaterals, square numbers, square units of area, magic squares, and of how 4 appears in literature, history, nature, art, and science.

Sitomer, Mindel and Harry Sitomer. *How Did Numbers Begin?* New York: Thomas Y. Crowell, 1976.
Clarifying several concepts which are central to our perception of number, the author helps pupils think about the preponderance of numbers in our lives and shows how the use of numbers likely developed.

Srivastava, Jane Jonas. *Number Families.* New York: Thomas Y. Crowell, 1979.
Family members often have at least one thing in common, and the same is true for numbers. This book visits odd and even numbers, numeral shapes, multiplication, multiples, factors, and prime numbers, as well as square and triangular numbers. The reader is guided through hands-on activities that make these concepts seem simple.

St. John, Glory. *How to Count Like a Martian.* New York: Henry Z. Walck, 1975.
   Counting methods and symbols used by ancient Egyptians, Babylonians, Mayas, Greeks, Chinese, and Hindus are explored in this book about different base systems. The author explains counting with an abacus and a computer, and leads children to decode a number base system, called the "Martian" system.

Whole Numbers—Suitable for Grades K-3

de Regniers, Beatrice Schenk. *So Many Cats.* New York: Clarion Books, 1985.
   This book depicts the story of a family with a lonely cat who doesn't stay lonely for long.

Hutchins, Pat. *The Doorbell Rang.* New York: Greenwillow Books, 1986.
   After their mother bakes a dozen fresh cookies for Victoria and Sam, a series of visitors arrive and the cookies have to be shared over and over again. The last visitor, however, is Grandma, who arrives with a huge tray of cookies she has baked.

Mathews, Louise. *Bunches and Bunches of Bunnies.* New York: Scholastic, 1978.
   Bunnies, depicted by colorful pictures and rhyming text, lead the reader to explore the multiplication facts up to 12 x 12.

Mathews, Louise. *The Great Take-Away.* New York: Dodd, Mead, 1980.
   With rhymes and colorful illustrations, this book shows examples of subtraction as a lazy pig steals from a his more industrious neighbors. Word problems are used throughout as the lazy pig proceeds to rob baby rattles, party gifts, necklaces, and pocketbooks until he is caught robbing the Piggy Bank.

McMillan, Bruce. *Counting Wildflowers.* New York: Lothrop, Lee & Shepard, 1986.
   Using beautiful photographs of wildflowers, this book illustrates the concepts of addition and subtraction up to twenty.

Owen, Annie. *Annie's One to Ten.* New York: Alfred A. Knopf, 1988.
   With illustrations, this book shows all of the different combinations of objects that can be grouped together to make ten.

Peterson, Esther Allen. *Penelope Gets Wheels.* New York: Crown, 1981.
   With her birthday money, Penelope decides to buy some "wheels" so she won't have to walk every place. Cars are not for her because they cost too much and she isn't old enough for a car, anyway. She shops for a bicycle, but they are also too expensive for her. Finally, she settles on roller skates and has a series of experiences that convince her that roller skates are the best wheels for kids.

Whole Numbers—Suitable for Grades 4-6

Anno, Masaichiro and Mitsumasa Anno. *Anno's Mysterious Multiplying Jar.* New York: Philomel Books, 1983.
   This story begins with one jar and the reader discovers there are 3,628,800 jars inside the one jar. An explanation of the concept of factorials clarifies the situation.

Asch, Frank. *Popcorn.* New York: Parents Magazine, 1976.
   Left alone when his parents go to a Halloween party, Sam throws a party for his friends. Having asked each guest to bring a snack, Sam discovers that everyone has brought popcorn! The kids pop all of the popcorn, filling the whole house, then they become sick after eating it all. And to top it all off, Sam's parents bring him a surprise when they come home—a box of popcorn!

Cole, Joanna. *The Magic School Bus Lost in the Solar System.* New York: Scholastic, 1990.
   The weirdest teacher in the school, Mrs. Frizzle, takes her class on a field trip to the planetarium, but they end up cruising on a bus through the solar system.

Merrill, Jean. *The Toothpaste Millionaire.* Boston: Houghton Mifflin, 1972.
   Because sixth grader Rufus Mayflower became disgruntled about the price of toothpaste, he begins making toothpaste in his kitchen. Soon, Rufus and his friend, Kate, are into a very profitable business.

Schwartz, David M. *How Much Is a Million?* New York: Lothrop, Lee & Shepard Books, 1985.
   Wonderfully illustrated with detailed illustrations, the reader is taken on a journey by Marvelosissimo, the mathematical magician, through an explanation of the concepts of million, billion, and trillion.

Viorst, Judith. *Alexander, Who Used to be Rich Last Sunday.* New York: Atheneum, 1976.
   Alexander only has bus tokens, but his brothers, Nicholas and Anthony, have money. After Alexander's grandparents give him one dollar, he wants to save it for a walkie-talkie, but, instead, he spends it quickly and winds up with—bus tokens.

Geometry and Spatial Sense—Suitable for Grades K-3

Anno, Mitsumasa. *Anno's Math Games III.* New York: Philomel Books, 1991.
   Mazes, topology, triangles, changing shapes, and left-to-right are among the many aspects of geometry and spatial relationships explored in this book.

Brown, Jeff. *Flat Stanley.* New York: Harper & Row, 1964. (Suitable for grades K-3.)
Stanley Lambchop is flattened like a pancake when a huge bulletin board falls on him. The story tells how Stanley's life changes because of his new dimensions, and how Arthur, his brother, calculates how to make Stanley his normal size again.

Dr. Seuss. *The Shape of Me and Other Stuff.* New York: Random House, 1988.
Children are stimulated to think about the shapes of real and imaginary things through rhymes and silhouettes.

Rogers, Paul. *The Shapes Game.* New York: Henry Holt, 1989.
Through rhyming text and colorful illustrations, children are led to identify circles, triangles, squares, ovals, crescents, rectangles, spirals, and diamonds.

Sitomer, Mindel and Harry Sitomer. *What Is Symmetry?* New York: Thomas Y. Crowell, 1970.
In ways that are easy for children to understand, this book explores the concepts of line, point, and plane symmetry.

Tompert, Ann. *Grandfather Tang's Story.* New York: Crown, 1990.
This book tells the story of Little Soo and Grandfather Tang spending an afternoon in the backyard making different shapes with tangram pieces. Using his pieces, Grandfather Tang tells a story about the fox fairies who, according to Chinese folklore, are able to change their shapes.

Geometry and Spatial Sense—Books Suitable for Grades 4-6

Froman, Robert. *Rubber Bands, Baseballs, and Doughnuts.* New York: Thomas Y. Crowell, 1972.
In this book, the branch of mathematics called topology is simplified through experiments with everyday objects. Topology is defined in this book as "the study of what does not change when line segments and different kinds of shapes are scrunched up, twisted around, or distorted in other ways." Examples of professions which use topology include astronomy, cartography, and computer design.

Goldreich, Gloria and Esther Goldreich. *What Can She Be? An Architect.* New York: Lothrop, Lee & Shepard, 1974.
This books follows the architect Susan Brody through several projects, leading pupils to discover the kinds of information and skills that architects must acquire.

Laithwaite, Eric. Shape: *The Purpose of Forms.* New York: Franklin Watts, 1986.
Addressing the properties and purposes of shapes, how shapes are made and used, and the concept of symmetry, this book motivates pupils to think about

the various shapes they encounter in the natural and manmade world.

Sitomer, Mindel and Harry Sitomer. *Spirals.* New York: Thomas Y. Crowell, 1974.
    This book includes activities to help children understand the geometric form called spirals. The differences between spirals and circles are clarified, and the book discusses spirals in both nature and manmade items.

Small, David. *Paper John.* New York: Farrar, Straus & Giroux, 1987.
    Paper John, a stranger, comes to town, builds himself a paper house, and earns a living selling paper items that he makes. While fishing one day, Paper John rescues a demon who tries both his patience and his ingenuity.

Snape, Juliet and Charles Snape. *The Boy with Square Eyes.* New York: Prentice-Hall Books for Young Readers, 1987.
    Because he watches so much television, Charlie's eyes become square-shaped. Everything looks square until he finds a way to get his eyes back to their normal shape.

Measurement—Books Suitable for Grades K-3

Carle, Eric. *The Grouchy Ladybug.* New York: Thomas Y. Crowell, 1977.
    A grouchy ladybug and a friendly ladybug meet on the same leaf to share a meal of aphids, but the grouchy ladybug doesn't want to share and threatens the friendly ladybug. Finally, the grouchy ladybug flies off and ends up challenging an assortment of other animals (even a whale), but ends up backing down each time. In the end, the grouchy ladybug returns to the same leaf where she began and shares with the friendly ladybug after all.

Heit, Robert. *The Day That Monday Ran Away.* New York: The Lion Press, 1969.
    Feeling unappreciated, Monday decides to run away. The problems that this creates are numerous, and finally Monday is found and persuaded to return.

Lionni, Leo. *Pezzettino.* New York: Pantheon, 1975.
    This book is about Pezzettino who was so small that he thought he must be a missing piece of somebody else. After voyaging through a series of "am I a piece of you?" experiences, Pezzettino finally realizes that he is himself.

Myller, Rolf. *How Big Is A Foot?* New York: Atheneum, 1962.
    A delightful story about a king and a queen who had everything. The King was perplexed as to what present to get his wife for her birthday. He finally decided to have her a bed made, "which was an excellent idea since beds hadn't been invented yet." The King instructed the apprentice carpenter to build the Queen's bed three feet wide and six feet long. But the bed was too small for the Queen, and the carpenter was banished to the dungeon for the mistake. Finally, the apprentice carpenter figures out the problem, tells the

King his solution, and then builds a bed to fit the Queen.

Schlein, Miriam. *Heavy Is A Hippopotamus.* New York: William R. Scott, 1954.
This book explores the concepts of heavy and light, and demonstrates the need for standardized measures.

Srivastava, Jane Jonas. *Space, Shapes, and Sizes.* New York: Thomas Y. Crowell, 1980.
This book uses bears, dogs, lions, pigs, and rabbits to illustrate simple experiments for children to help them learn more about volume.

Measurement—Books Suitable for Grades 4-6

Cole, Joanna. *The Magic School Bus Inside the Earth.* New York: Scholastic, 1987.
Because most of her class didn't bring a rock to school as directed, Mrs. Frizzle decides to take the class on a field trip to collect rocks. Their field trip turns into an exciting adventure when they travel to the center of the Earth and back home again.

Cole, Joanna. *The Magic School Bus Inside the Human Body.* New York: Scholastic, 1990.
On a field trip to the museum, Mrs. Frizzle's class stops at a park for lunch. Arnold, however, misses the bus and remains at the picnic table eating his Cheesie-Weesies. But instead of driving away, the bus becomes very tiny, slips into Arnold's snack bag, and Arnold swallows the bus without noticing what he has done. The class takes a trip on the bus through the human body.

Oakley, Graham. *Diary of a Church Mouse.* New York: Atheneum, 1987.
This book is all about a church mouse named Humphrey who kept a diary about his life. This book contains one year in the life of Humphrey.

Paine, Penelope Colville. *Time for Horatio.* Santa Barbara, CA: Advocacy Press, 1990.
A stray kitten named Horatio is taken aboard the pleasure steamer out of London, the Victorious. Many of the tourists are mean to Horatio, and he misunderstands Greenwich Mean Time, thinking that is what is making the people mean. Oliver, the boy who found Horatio and brought him aboard, has a dream in which Horatio stops Big Ben in order to stop mean time.

Verne, Jules. *Around the World in Eighty Days.* New York: William Morrow, 1988.
An eccentric Englishman named Phileas Fogg wagers £20,000 that he can travel all the way around the world in 80 days in the year 1872. In spite of numerous diversions, delays, and adventures, Phileas Fogg makes the trip and returns home just at the last moment to win the wager.

Wolkstein, Diane. *8,000 Stones.* Garden City, NY: Doubleday, 1972
    The Satrap of India sends Ts'ao Ts'ao, the Most Supreme Governor of China, an elephant who is ten feet tall. No one in the kingdom has ever seen an elephant, and Ts'ao Ts'ao directs his advisors to find out how much the elephant weighs. No one can think of a way to weigh the elephant until P'ei, Ts'ao Ts'ao's son, comes up with a solution.

Statistics and Probability—Books Suitable for Grades K-3
Linn, Charles F. *Probability.* New York: Thomas Y. Crowell, 1972.
    The concept of probability is explained as two mice lead children through some simple experiments. The book also shows ways to arrange the data from the experiments so they can be utilized for making predictions.

Munsch, Robert. *Moira's Birthday.* Toronto: Annick Press, 1989.
    When Moira invites "grade 1, grade 2, grade 3, grade 4, grade 5, aaaaand kindergarten" to her birthday party, her parents expect six children for the party. But 200 show up for the party! Moira orders 200 birthday cakes and 200 pizzas, but the bakery and the pizza parlor can only deliver ten at first. Moira offers to share the 200 presents she received if her guests will help her clean up. After everyone has gone, the rest of the pizzas and birthday cakes arrive. Moira's solution is to have another party!

Sachs, Marilyn. *Fleet-Footed Florence.* Garden City, NY: Doubleday, 1981.
    An ex-baseball great hopes his three sons grow up to be great, too. But it is his daughter, Florence, who becomes the fastest runner in the East, West, North, and South. She sets so many records there was no book big enough to hold them all.

Srivastava, Jane Jonas. *Averages.* New York: Thomas Y. Crowell, 1975.
    In an easy-to-understand format, this books explains concepts like the median, the mode, and the arithmetical mean. Lots of examples help children realize that the type of problem you want to solve determines which type of average you use.

Srivastava, Jane Jonas. *Statistics.* New York: Thomas Y. Crowell, 1973.
    This book explains how statistics are found primarily through counting, how statistics are gathered, and how graphs are used to communicate statistics.

Statistics and Probability—Books Suitable for Grades 4-6
Cushman, Jean. *Do You Wanna Bet?* New York: Clarion Books, 1991.
    In this book, two boys find that chance plays a big part in their lives as they learn about probability with baseball statistics, PTA carnivals, baby sisters, coin tosses, weather forecasts, and secret codes.

James, Elizabeth and Carol Barkin. *What Do You Mean by "Average"?* New York: Lothrop, Lee & Shepard, 1978.
> In running for president of the student council, Jill decides that the best strategy is to prove that she's a very average person. Jill and her campaign staff utilize different kinds of averages in the process of proving that Jill is the most average person in the school.

MacGregor, Ellen and Dora Pantell. *Miss Pickerell and the Weather Satellite.* New York: McGraw-Hill, 1971.
> While heavy rains, thunder, lightning, and high winds pound Square Toe Mountain, the weather satellites and computers at the weather station are forecasting sunshine and clear skies. To prevent the county from flooding, the dam must be opened, but the dam can only be opened when the weather station predicts heavy rain. To open the dam and save the county from flooding, Miss Pickerell must make a trip to the weather satellite itself.

Mori, Tuyosi. *Socrates and the Three Little Pigs.* New York: Philomel Books, 1986.
> Socrates, a philosopher wolf, spies on three little pigs playing in the meadow. Socrates' wife asks him to catch one of the pigs for dinner. There are five houses in which the pigs live, and Socrates turns to his mathematician friend, Pythagoras (a frog) for help in determining the best way to find one pig by itself. Socrates and Pythagoras explore various combinations and variations involved in solving this problem, but by the time they arrive at a solution, it is daylight and they abandon their plan. Instead, they decide to join the pigs at play in the meadow.

Roy, Ron. *Million Dollar Jeans.* New York: E. P. Dutton, 1983.
> Two ten-year-old boys have a roundabout hunt and many adventures as they seek a lost lottery ticket worth a million dollars.

Fractions and Decimals—Books Suitable for Grades K-3

Dennis, J. Richard. *Fractions Are Parts of Things.* New York: Thomas Y. Crowell, 1971.
> The simple fractions of halves, thirds, and fourths are explored with activities and illustrations.

Hoban, Lillian. *Arthur's Funny Money.* New York: Harper & Row, 1981.
> Along with his little sister, Violet, Arthur sets up a bike-washing business to get five dollars to buy a "Far Out Frisbees" T-shirt and matching cap. After their first customers come, a dog eats the soap out of the box and Arthur and Violet set off for the store to get more soap. On the way back, they see the T-shirt and cap displayed in a store window with a sign discounting the price. Arthur finds that he has enough money to buy the T-shirt and matching cap and have eighteen cents left to buy licorice twists for himself and Violet.

Howe, James. *Harold & Chester in Hot Fudge.* New York: Morrow Junior Books, 1990.
   Chester the cat, Harold and Howie the dogs, and Bunnicula the rabbit are the four pets of the Monroe family. When the fudge that Mr. Monroe makes for the Library Bake Sale appears to have been stolen, the pets work to clear up the mystery and make sure the fudge gets to the bake sale.

Maestro, Betsy and Giulio Maestro. *Dollars and Cents for Harriet.* New York: Crown, 1988.
   Harriet the elephant wants a toy that costs five dollars. When she looks in her piggybank, she finds that she has 100 pennies. The book shows how Harriet earns the other four dollars in different denominations of coins.

Mathews, Louise. *Gator Pie.* New York: Dodd, Mead, 1979.
   Delighted at finding a chocolate marshmallow pie on a picnic table near the edge of the swamp, two alligators, Alice and Alvin, decide to split the pie into halves and eat it. But as they begin, another alligator tramps up and demands some of the pie. More gators come demanding pie, and finally Alice has to cut the pie into 100 pieces. As Alvin climbs onto the table and tells the alligators to pick up a piece of pie, he alerts them that all pieces may not be the same size. When the alligators hear this, bedlam breaks out and the gators begin fighting. While the fight is going on, Alice and Alvin whisk the pie away and each enjoy eating 50 pieces of chocolate marshmallow pie.

Fractions and Decimals—Books Suitable for Grades 4-6

Adams, Laurie and Allison Coudert. *Alice and the Boa Constrictor.* Boston: Houghton Mifflin, 1983.
   While studying reptiles in the fourth grade at Miss Barton's School for Girls, Alice Whipple decides she wants to buy her own boa constrictor. Thinking that Alice could never earn enough money to buy the snake, her father reluctantly agrees to the deal. Alice, however, displays some very clever and creative ways to earn the money!

Cole, Joanna. *The Magic School Bus at the Waterworks.* New York: Scholastic, 1986.
   On a field trip to the waterworks, Mrs. Frizzle's class has an amazing adventure when the school bus evaporates into the sky like water does. The children have a variety of adventures that help explain the water cycle, including being encased in water droplets dripping from a faucet in the girls' bathroom of the school!

Conford, Ellen. *What's Cooking, Jenny Archer?* Boston: Little, Brown, 1989.
   Her dislike for school lunches leads Jenny Archer to decide to make her own

lunches. Soon, Jenny's friends offer her their lunch money to make their lunches as well. Calculating her anticipated profit, Jenny quickly agrees, but she didn't take into consideration "finicky" eaters, hungry dogs, and the expenses involved!

Klevin, Jill Ross. *The Turtle Street Trading Co.* New York: Delacorte Press, 1982.
The Turtles is the name of a secret society of four twelve-year-olds who have nearly abandoned their idea of a trip to Disneyland because they have had little success in raising money. But instead, they start a business called The Turtle Street Trading Co. where they salvage old junk like used toys, tapes, records, posters, models, and T-shirts for which other kids can trade their old things. The business gets off to a rocky start, but then becomes a tremendous success.

Levitin, Sonia. *Jason and the Money Tree.* New York: Harcourt Brace Jovanovich, 1974.
After receiving a box of treasures (including a ten-dollar bill) when his grandfather dies, eleven year old Jason Galloway remembers a favorite adage of his grandfather, "Nothing is impossible." With that in mind, Jason plants the ten-dollar bill in the backyard, and sure enough, a money tree sprouts! But the money tree is not quite the good fortune that it seems, and Jason confronts many unanticipated dilemmas before the story ends.

Schwartz, David M. *If You Made a Million.* New York: Lothrop, Lee & Shepard, 1989.
Clever illustrations and Marvelosissimo, the mathematical magician, help children understand concepts of equivalence, such as "one dime has the same value as two nickels or ten pennies." Marvelosissimo goes all the way to one million dollars in his illustrations, and he also explains about interest, writing checks, and the buying power of various amounts of money.

Patterns and Relationships—Books Suitable for Grades K-3

Bayer, Jane. *A, My Name Is Alice.* New York: Dial Books for Young Readers, 1984.
Using an alliterative pattern, this book is based on a game that takes the children through the alphabet from A to Z. For example, "A, My name is Alice and my husband's name is Alex. We come from Alaska and we sell Ants."

Charosh, Mannis. *Number Ideas Through Pictures.* New York: Thomas Y. Crowell, 1974.
The concepts of odd and even numbers, triangular numbers, and square numbers are clarified by the use of clear illustrations and simple explanations using manipulatives.

Ernst, Lisa Campbell. *Sam Johnson and the Blue Ribbon Quilt.* New York: Lothrop, Lee & Shepard, 1983.
> While his wife is away, farmer Sam Johnson learns to quilt. But when he tries to join his wife's quilting club, the women laugh at him, so Sam starts the Rosedale Men's Quilting Club. The two groups become competitors at the county fair quilting contest, but end up working together for an award.

Rinkoff, Barbara. *The Case of the Stolen Code Book.* New York: Crown, 1971.
> Secret Agents Club members Alex, John, Holly, and Winnie leave their code book in the yard. New neighbor, Bob, finds the code book and leaves clues and codes to be followed in order to retrieve the missing code book. The group is so impressed that they ask Bob to join their club.

Zolotow, Charlotte. *Some Things Go Together.* New York: Thomas Y. Crowell, 1969.
> Various ideas, objects, and actions that are connected in some way are explored through rhyming couplets.

Patterns and Relationships—Books Suitable for Grades 4-6

Robertson, Joanne and Laszlo Gal. *Sea Witches.* New York: Dial Books for Young Readers, 1991.
> Haiku verse tells Scottish tale of how ghost sea witches turn uncrushed egg shells into boats and search the seas for ships to sink.

Singer, Marilyn. *A Clue in Code.* New York: Harper & Row, 1985.
> Sam and Dave Bean, twins in Mrs. Corfein's classroom, work to solve the mystery of the missing class trip money. Aided by a paper airplane with a coded message and their friend, Rita O'Toole, they find the person who took the money.

Terban, Marvin. *Too Hot to Hoot.* New York: Clarion, 1985.
> Palindromes are words, phrases, sentences, or numbers that are written the same forwards and backwards. This book deals with language and numerical palindromes and their patterns.

Thompson, Dr. David. *Visual Magic.* New York: Dial Books, 1991.
> This book explores how our visual system interprets the world around us, but how sometimes we can be fooled.

Weiss, Malcolm E. *Solomon Grundy, Born on Oneday.* New York: Thomas Y. Crowell, 1977.
> "Solomon Grundy, born on Monday, christened on Tuesday, married on Wednesday, took sick on Thursday, buried on Sunday." Did this nursery rhyme really all happen in one week, or can it be explained in another way?

… Activities for Elementary School Mathematics

STANDARDS FOR SCHOOL MATHEMATICS[1]

The National Council of Teachers of Mathematics defines a mathematics standard as a statement that can be used to judge the quality of a mathematics curriculum. In 1986, the Board of Directors of the National Council of Teachers of Mathematics established the Commission on Standards for School Mathematics as a means to help improve the quality of school mathematics. The product of the commission's efforts is a document containing a set of standards for mathematics curricula in North American schools (K-12) and for evaluating the quality of both the curriculum and pupil achievement. As school staffs, school districts, states, provinces, and other groups propose solutions to mathematics curricular problems, these standards may be used as criteria against which their ideas can be judged.

The standards document is designed to establish a broad framework to guide reform in school mathematics well into the future. The document provides a direction and focus for specific content areas as well as a general philosophy of mathematics education. It gives a vision of what the mathematics curriculum should include in terms of content priority and content emphasis. Those working to improve the quality of school mathematics are challenged to work collaboratively to use these curriculum and evaluation standards as the basis for change so that the teaching and learning of mathematics in schools is improved.

CURRICULUM STANDARDS FOR GRADES K-4

Overview
Thirteen curriculum standards are presented for grades K-4:
1. Mathematics as Problem Solving
2. Mathematics as Communication
3. Mathematics as Reasoning
4. Mathematical Connections
5. Estimation
6. Number Sense and Numeration
7. Concepts of Whole Number Operations
8. Whole Number Computation
9. Geometry and Spatial Sense
10. Measurement
11. Statistics and Probability
12. Fractions and Decimals
13. Patterns and Relationships

---

[1] The National Council of Teachers of Mathematics. Used by permission.

Children and Mathematics: Implications for the K-4 Curriculum

An appropriate curriculum for young children that reflects the Standards' overall goals must do the following:

1. Address the relationship between young children and mathematics. Children enter kindergarten with considerable mathematical experience, a partial understanding of many concepts, and some important skills, including counting. Nonetheless, it takes careful planning to create a curriculum that capitalizes on children's intuitive insights and language in selecting and teaching mathematical ideas and skills. It is clear that children's intellectual, social, and emotional development should guide the kind of mathematical experiences they should have in light of the overall goals for learning mathematics. The notion of a developmentally appropriate curriculum is an important one. A developmentally appropriate curriculum encourages the exploration of a wide variety of mathematical ideas in such a way that children retain their enjoyment of, and curiosity about, mathematics. It incorporates real-world contexts, children's experiences, and children's language in developing ideas. It recognizes that children need considerable time to construct sound understandings and develop the ability to reason and communicate mathematically. It looks beyond what children appear to know to determine how they think about ideas. It provides repeated contact with important ideas in varying contexts throughout the year and from year to year. Programs that provide limited developmental work, that emphasize symbol manipulation and computational rules, and that rely heavily on paper-and-pencil worksheets do not fit the natural learning patterns of children and do not contribute to important aspects of children's mathematical development.

2. Recognize the importance of the qualitative dimensions of children's learning. The mathematical ideas that children acquire in grades K-4 form the basis for all further study of mathematics. Although quantitative considerations have frequently dominated discussions in recent years, qualitative considerations have greater significance. Thus, how well children come to understand mathematical ideas is far more important than how many skills they acquire. The success with which programs at later grade levels achieve their goals depends largely on the quality of the foundation that is established during the first five years of school.

3. Build beliefs about what mathematics is, about what it means to know and do mathematics, and about children's views of themselves as mathematics learners. The beliefs that young children form influence not only their thinking and performance during this time but also their attitude and decisions about studying mathematics in later years. Beliefs also become more resistant to change as children grow older. Thus, affective dimensions of learning play a significant role in, and must influence, curriculum and instruction.

Activities for Elementary School Mathematics

Basic Assumptions

Several basic assumptions governed the selection and shaping of the K-4 standards:

1. The K-4 curriculum should be conceptually oriented. The view that the K-4 curriculum should emphasize the development of mathematical understandings and relationships is reflected in the discussions about the content and emphasis of the curriculum. A conceptual approach enables children to acquire clear and stable concepts by constructing meanings in the context of physical situations and allows mathematical abstractions to emerge from empirical experience. A strong conceptual framework also provides anchoring for skill acquisition. Skills can be acquired in ways that make sense to children and in ways that result in more effective learning. A strong emphasis on mathematical concepts and understandings also supports the development of problem solving.

Emphasizing mathematical concepts and relationships means devoting substantial time to the development of understandings. It also means relating this knowledge to the learning of skills by establishing relationships between the conceptual and procedural aspects of tasks. The time required to build an adequate conceptual base should cause educators to rethink when children are expected to demonstrate a mastery of complex skills. A conceptually oriented curriculum is consistent with the overall curricular goals in this report and can result in programs that are better balanced, more dynamic, and more appropriate to the intellectual needs and abilities of children.

2. The K-4 curriculum should actively involve children in doing mathematics. Young children are active individuals who construct, modify, and integrate ideas by interacting with the physical world, materials, and other children. Given these facts, it is clear that the learning of mathematics must be an active process. Throughout the Standards, such verbs as *explore, justify, represent, solve, construct, discuss, use, investigate, describe, develop,* and *predict* are used to convey this active physical and mental involvement of children in learning the content of the curriculum.

The importance of active learning by children has many implications for mathematics education. Teachers need to create an environment that encourages children to explore, develop, test, discuss, and apply ideas. They need to listen carefully to children and to guide the development of their ideas. They need to make extensive and thoughtful use of physical materials to foster the learning of abstract ideas.

K-4 classrooms need to be equipped with a wide variety of physical materials and supplies. Classrooms should have ample quantities of such materials as counters; interlocking cubes; connecting links; base-ten, attribute, and pattern blocks; tiles; geometric models; rulers; spinners; colored rods; geoboards; balances; fraction pieces; and graph, grid, and dot paper. Simple household objects, such as buttons, dried beans, shells, egg cartons, and milk cartons also can be used.

3. The K-4 curriculum should emphasize the development of children's mathematical thinking and reasoning abilities. An individual's future uses and needs for mathematics make the ability to think, reason, and solve problems a primary goal for the study of mathematics. Thus, the curriculum must take seriously the goal of instilling in pupils a sense of confidence in their ability to think and communicate mathematically, to solve problems, to demonstrate flexibility in working with mathematical ideas and problems, to make appropriate decisions in selecting strategies and techniques, to recognize familiar mathematical structures in unfamiliar settings, to detect patterns, and to analyze data. The K-4 standards reflect the view that mathematics instruction should promote these abilities so that pupils understand that knowledge is empowering and that individual pieces of content are all related to this broader perspective.

Developing these characteristics in children requires that schools build appropriate reasoning and problem-solving experiences into the curriculum from the outset. Further, this goal needs to influence the way mathematics is taught and the way pupils encounter and apply mathematics throughout their education.

4. The K-4 curriculum should emphasize the application of mathematics. If children are to view mathematics as a practical, useful subject, they must understand that it can be applied to a wide variety of real-world problems and phenomena. Even though most mathematical ideas in the K-4 curriculum arise from the everyday world, they must be regularly applied to real-world situations. Children also need to understand that mathematics is an integral part of real-world situations and activities in other curricular areas. The mathematical aspects of that work should be highlighted.

Learning mathematics has a purpose. At the K-4 level, one major purpose is helping children understand and interpret their world and solve problems that occur in it. Children learn computation to solve problems; they learn to measure because measurement helps them answer questions about how much, how big, how long, and so on; and they learn to collect and organize data because doing so permits them to answer other questions. By applying mathematics, they learn to appreciate the power of mathematics.

5. The K-4 curriculum should include a broad range of content. To become mathematically literate, pupils must know more than arithmetic. They must possess a knowledge of such important branches of mathematics as measurement, geometry, statistics, probability, and algebra. These increasingly important and useful branches of mathematics have significant and growing applications in many disciplines and occupations.

The curriculum at all levels needs to place substantial emphasis on these branches of mathematics. Mathematical ideas grow and expand as children work with them throughout the curriculum. The informal approach at this level establishes the foundation for further study and permits children to acquire additional knowledge they will need. These topics are highly appropriate for young learners because they make important contributions to children's mathematical development and help them see the usefulness of mathematics. They also provide productive, intriguing activities and applications.

The inclusion of a broad range of content in the curriculum also allows children to see the interrelated nature of mathematical knowledge. When teachers take advantage of the opportunity to relate one mathematical idea to others and to other areas of the curriculum, as will be described in Standard 4, children acquire broader notions about the interconnectedness of mathematics and its relationships to other fields. The curriculum should enable all children to do a substantial amount of work in each of these topics at each grade level.

6. The K-4 curriculum should make appropriate and ongoing use of calculators and computers. Calculators must be accepted at the K-4 level as valuable tools for learning mathematics. Calculators enable children to explore number ideas and patterns, to have valuable concept-development experiences, to focus on problem-solving processes, and to investigate realistic applications. The thoughtful use of calculators can increase the quality of the curriculum as well as the quality of children's learning.

Calculators do not replace the need to learn basic facts, to compute mentally, or to do reasonable paper-and-pencil computation. Classroom experience indicates that young children take a commonsense view about calculators and recognize the importance of not relying on them when it is more appropriate to compute in other ways. The availability of calculators means, however, that educators must develop a broader view of the various ways computation can be carried out and must place less emphasis on complex paper-and-pencil computation. Calculators also highlight the importance of teaching children to recognize whether computed results are reasonable.

The power of computers also needs to be used in contemporary mathematics programs. Computer languages that are geometric in nature help young children become familiar with important geometric ideas. Computer simulations of mathematical ideas, such as modeling the renaming of numbers, are an important aid in helping children identify the key features of the mathematics. Many software programs provide interesting problem-solving situations and applications.

The thoughtful and creative use of technology can greatly improve both the quality of the curriculum and the quality of children's learning. Integrating calculators and computers into school mathematics programs is critical in meeting the goals of a redefined curriculum.

SUMMARY OF CHANGES IN CONTENT AND EMPHASIS IN K--4 MATHEMATICS

GIVE MORE ATTENTION TO:

NUMBER
- Number sense
- Place-value concepts
- Meaning of fractions and decimals
- Estimation of quantities

OPERATIONS AND COMPUTATION
- Meaning of operations
- Operation sense
- Mental computation
- Estimation and the reasonableness of answers
- Selection of an appropriate computational method
- Use of calculators for complex computation
- Thinking strategies for basic facts

GEOMETRY AND MEASUREMENT
- Properties of geometric figures
- Geometric relationships
- Spatial sense
- Process of measuring
- Concepts related to units of measurement
- Actual measuring
- Estimation of measurements
- Use of measurement and geometry ideas throughout the curriculum

PROBABILITY AND STATISTICS
- Collection and organization of data
- Exploration of chance

PATTERNS AND RELATIONSHIPS
- Pattern recognition and description
- Use of variables to express relationships

PROBLEM SOLVING
- Word problems with a variety of structures
- Use of everyday problems
- Applications
- Study of patterns and relationships
- Problem-solving strategies

INSTRUCTIONAL PRACTICES
- Use of manipulative materials
- Cooperative work
- Discussion of mathematics
- Questioning
- Justification of thinking

Writing about mathematics
Problem-solving approach to instruction
Content integration
Use of calculators and computers

GIVE LESS ATTENTION TO:
NUMBER
Early attention to reading, writing, and ordering numbers symbolically
OPERATIONS AND COMPUTATION
Complex paper-and-pencil computations
Isolated treatment of paper-and-pencil computations
Addition and subtraction without renaming
Isolated treatment of division facts
Long division
Long division without remainders
Paper-and-pencil fraction computation
Use of rounding to estimate
GEOMETRY AND MEASUREMENT
Primary focus on naming geometric figures
Memorization of equivalencies between units of measurement
PROBLEM SOLVING
Use of clue words to determine which operation to use
INSTRUCTIONAL PRACTICES
Rote practice
Rote memorization of rules
One answer and one method
Use of worksheets
Written practice
Teaching by telling

STANDARD 1: MATHEMATICS AS PROBLEM SOLVING
In grades K-4, the study of mathematics should emphasize problem solving so that pupils can--
use problem-solving approaches to investigate and understand mathematical content;
formulate problems from everyday and mathematical situations;
develop and apply strategies to solve a wide variety of problems;
verify and interpret results with respect to the original problem;
acquire confidence in using mathematics meaningfully.

Focus

Problem solving should be the central focus of the mathematics curriculum. As such, it is a primary goal of all mathematics instruction and an integral part of all mathematical activity. Problem solving is not a distinct topic but a process that should permeate the entire program and provide the context in which concepts and skills can be learned.

This standard emphasizes a comprehensive and rich approach to problem solving in a classroom climate that encourages and supports problem-solving efforts. Ideally, pupils should share their thinking and approaches with other pupils and with teachers, and they should learn several ways of representing problems and strategies for solving them. In addition, they should learn to value the process of solving problems as much as they value the solutions. Pupils should have many experiences in creating problems from real-world activities, from organized data, and from equations.

STANDARD 2: MATHEMATICS AS COMMUNICATION

In grades K-4, the study of mathematics should include numerous opportunities for communication so that pupils can--

> relate physical materials, pictures, and diagrams to mathematical ideas;
> reflect on and clarify their thinking about mathematical ideas and situations;
> relate their everyday language to mathematical language and symbols;
> realize that representing, discussing, reading, writing, and listening to
>   mathematics are a vital part of learning and using mathematics.

Focus

Mathematics can be thought of as a language that must be meaningful if pupils are to communicate mathematically and apply mathematics productively. Communication plays an important role in helping children construct links between their informal, intuitive notions and the abstract language and symbolism of mathematics; it also plays a key role in helping children make important connections among physical, pictorial, graphic, symbolic, verbal, and mental representations of mathematical ideas. When children see that one representation, such as an equation, can describe many situations, they begin to understand the power of mathematics; when they realize that some ways of representing a problem are more helpful than others, they begin to understand the flexibility and usefulness of mathematics.

Young children learn language through verbal communication; it is important, therefore, to provide opportunities for them to "talk mathematics." Interacting with classmates helps children construct knowledge, learn other ways to think about ideas, and clarify their own thinking. Writing about mathematics, such as describing how a problem was solved, also helps pupils clarify their thinking and develop deeper understanding. Reading children's literature about mathematics, and eventually text material, also is an important aspect of communication that needs more emphasis in the K-4 curriculum.

STANDARD 3: MATHEMATICS AS REASONING
In grades K-4, the study of mathematics should emphasize reasoning so that pupils can--
> draw logical conclusions about mathematics;
> use models, known facts, properties, and relationships to explain their thinking;
> justify their answers and solution processes;
> use patterns and relationships to analyze mathematical situations;
> believe that mathematics makes sense.

Focus

A major goal of mathematics instruction is to help children develop the belief that they have the power to do mathematics and that they have control over their own success or failure. This autonomy develops as children gain confidence in their ability to reason and justify their thinking. It grows as children learn that mathematics is not simply memorizing rules and procedures but that mathematics makes sense, is logical, and is enjoyable. A classroom that values reasoning also values communicating and problem solving, all of which are components of the broad goals of the entire elementary school curriculum.

A climate should be established in the classroom that places critical thinking at the heart of instruction. Both teachers' and children's statements should be open to question, reaction, and elaboration from others in the classroom. Such a climate depends on all members of the class expressing genuine respect and support for one another's ideas. Children need to know that being able to explain and justify their thinking is important and that how a problem is solved is as important as its answer. This mind-set is established when children have opportunities to apply their reasoning skills and when justifying one's thinking is an expected component of problem discussions.

STANDARD 4: MATHEMATICAL CONNECTIONS
In grades K-4, the study of mathematics should include opportunities to make connections so that pupils can--
> link conceptual and procedural knowledge;
> relate various representations of concepts or procedures to one another;
> recognize relationships among different topics in mathematics;
> use mathematics in other curriculum areas;
> use mathematics in their daily lives.

Focus

This standard's purpose is to help children see how mathematical ideas are related. The mathematics curriculum is generally viewed as consisting of several discrete strands. As a result, computation, geometry, measurement, and problem solving tend to be taught in isolation. It is important that children connect ideas both among and within areas of mathematics. Without such connections, children must learn and remember too many isolated concepts and skills rather than recognizing general principles relevant to several areas. When mathematical ideas are also connected to everyday experiences, both in and out of school, children become aware of the usefulness of mathematics.

STANDARD 5: ESTIMATION

In grades K-4, the curriculum should include estimation so pupils can--
- explore estimation strategies;
- recognize when an estimate is appropriate;
- determine the reasonableness of results;
- apply estimation in working with quantities, measurement, computation, and problem solving.

Focus

Estimation presents pupils with another dimension of mathematics; terms such as about, near, closer to, between, and a little less than illustrate that mathematics involves more than exactness. Estimation interacts with number sense and spatial sense to help children develop insights into concepts and procedures, flexibility in working with numbers and measurements, and an awareness of reasonable results. Estimation skills and understanding enhance the abilities of children to deal with everyday quantitative situations.

From children's earliest experiences with mathematics, estimation needs to be an ongoing part of their study of numbers, computation, and measurement. It is important that children learn a variety of methods of estimating, such as the front-end strategy for computation and the chunking procedure for measurement. They also need to develop reasoning, judgment, and decision-making skills in using estimation.

STANDARD 6: NUMBER SENSE AND NUMERATION

In grades K-4, the mathematics curriculum should include whole number concepts and skills so that pupils can--
- construct number meanings through real-world experiences and the use of physical materials;
- understand our numeration system by relating counting, grouping, and place-value concepts;
- develop number sense;

interpret the multiple uses of numbers encountered in the real world.

Focus

Children must understand numbers if they are to make sense of the ways numbers are used in their everyday world. They need to use numbers to quantify, to identify location, to identify a specific object in a collection, to name, and to measure. Furthermore, an understanding of place value is crucial for later work with number and computation.

Intuition about number relationships helps children make judgments about the reasonableness of computational results and of proposed solutions to numerical problems. Such intuition requires good number sense. Children with good number sense (1) have well-understood number meanings, (2) have developed multiple relationships among numbers, (3) recognize the relative magnitudes of numbers, (4) know the relative effect of operating on numbers, and (5) develop referents for measures of common objects and situations in their environments.

STANDARD 7: CONCEPTS OF WHOLE NUMBER OPERATIONS

In grades K-4, the mathematics curriculum should include concepts of addition, subtraction, multiplication, and division of whole numbers so that pupils can--
> develop meaning for the operations by modeling and discussing a rich variety of problem situations;
> relate the mathematical language and symbolism of operations to problem situations and informal language;
> recognize that a wide variety of problem structures can be represented by a single operation;
> develop operation sense.

Focus

Understanding the fundamental operations of addition, subtraction, multiplication, and division is central to knowing mathematics. One essential component of what it means to understand an operation is recognizing conditions in real-world situations that indicate that the operation would be useful in those situations. Other components include building an awareness of models and the properties of an operation, seeing relationships among operations, and acquiring insight into the effects of an operation on a pair of numbers. These four components are aspects of operation sense. Children with good operation sense are able to apply operations meaningfully and with flexibility. Operation sense interacts with number sense and enables pupils to make thoughtful decisions about the reasonableness of results. Furthermore, operation sense provides a framework for the conceptual development of mental and written computational procedures.

STANDARD 8: WHOLE NUMBER COMPUTATION

In grades K-4, the mathematics curriculum should develop whole number computation so that pupils can--

    model, explain, and develop reasonable proficiency with basic facts and algorithms;
    use a variety of mental computation and estimation techniques;
    use calculators in appropriate computational situations;
    select and use computation techniques appropriate to specific problems and determine whether the results are reasonable.

Focus

    The purpose of computation is to solve problems. Thus, although computation is important in mathematics and in daily life, our technological age requires us to rethink how computation is done today. Almost all complex computation today is done by calculators and computers. In many daily situations, answers are computed mentally or estimates are sufficient, and paper-and-pencil algorithms are useful when the computation is reasonably straightforward. This standard addresses the importance of teaching children a variety of ways to compute, as well as the usefulness of calculators in solving problems containing large numbers or requiring complex computations. Related to this goal is the necessity of having reasonable expectations for proficiency with paper-and-pencil computation. Clearly, paper-and-pencil computation cannot continue to dominate the curriculum or there will be insufficient time for children to learn other, more important mathematics they need now and in the future.

STANDARD 9: GEOMETRY AND SPATIAL SENSE

In grades K-4, the mathematics curriculum should include two-and three-dimensional geometry so that pupils can--

    describe, model, draw, and classify shapes;
    investigate and predict the results of combining, subdividing, and changing shapes;
    develop spatial sense;
    relate geometric ideas to number and measurement ideas;
    recognize and appreciate geometry in their world.

Focus

    Geometry is an important component of the K-4 mathematics curriculum because geometric knowledge, relationships, and insights are useful in everyday situations and are connected to other mathematical topics and school subjects. Geometry helps us represent and describe in an orderly manner the world in which we live. Children are naturally interested in geometry and find it intriguing and motivating; their spatial capabilities frequently exceed their numerical skills, and tapping these strengths can foster an interest in mathematics and improve number understandings and skills.

Spatial understandings are necessary for interpreting, understanding, and appreciating our inherently geometric world. Insights and intuitions about two- and three-dimensional shapes and their characteristics, the interrelationships of shapes, and the effects of changes to shapes are important aspects of spatial sense. Children who develop a strong sense of spatial relationships and who master the concepts and language of geometry are better prepared to learn number and measurement ideas, as well as other advanced mathematical topics.

STANDARD 10: MEASUREMENT

In grades K-4, the mathematics curriculum should include measurement so that pupils can--
- understand the attributes of length, capacity, weight, mass, area, volume, time, temperature, and angle;
- develop the process of measuring and concepts related to units of measurement;
- make and use estimates of measurement;
- make and use measurements in problem and everyday situations.

Focus

Measurement is of central importance to the curriculum because of its power to help children see that mathematics is useful in everyday life and to help them develop many mathematical concepts and skills. Measuring is a natural context in which to introduce the need for learning about fractions and decimals, and it encourages children to be actively involved in solving and discussing problems.

Instruction at the K-4 level emphasizes the importance of establishing a firm foundation in the basic underlying concepts and skills of measurement. Children need to understand the attribute to be measured as well as what it means to measure. Before they are capable of such understanding, they must first experience a variety of activities that focus on comparing objects directly, covering them with various units, and counting the units. Premature use of instruments or formulas leaves children without the understanding necessary for solving measurement problems.

STANDARD 11: STATISTICS AND PROBABILITY

In grades K-4, the mathematics curriculum should include experiences with data analysis and probability so that pupils can--
- collect, organize, and describe data;
- construct, read, and interpret displays of data;
- formulate and solve problems that involve collecting and analyzing data;
- explore concepts of chance.

Focus

Collecting, organizing, describing, displaying, and interpreting data, as well as making decisions and predictions on the basis of that information, are skills that are increasingly important in a society based on technology and communication. These processes are particularly appropriate for young children because they can be used to solve problems that often are inherently interesting, represent significant, applications of mathematics to practical questions, and offer rich opportunities for mathematical inquiry. The study of statistics and probability highlights the importance of questioning, conjecturing, and searching for relationships when formulating and solving real-world problems.

A spirit of investigation and exploration should permeate statistics instruction. Children's questions about the physical world can often be answered by collecting and analyzing data. After generating questions, they decide what information is appropriate and how it can be collected, displayed, and interpreted to answer their questions. The analysis and evaluation that occur as children attempt to draw conclusions about the original problem often lead to new conjectures and productive investigations. This entire process broadens children's views of mathematics and its usefulness.

STANDARD 12: FRACTIONS AND DECIMALS

In grades K-4, the mathematics curriculum should include fractions and decimals so that pupils can--
> develop concepts of fractions, mixed numbers, and decimals;
> develop number sense for fractions and decimals;
> use models to relate fractions to decimals and to find equivalent fractions;
> use models to explore operations on fractions and decimals;
> apply fractions and decimals to problem situations.

Focus

Fractions and decimals represent a significant extension of children's knowledge about numbers. When children possess a sound understanding of fraction and decimal concepts, they can use this knowledge to describe real-world phenomena and apply it to problems involving measurement, probability, and statistics. An understanding of fractions and decimals broadens pupils' awareness of the usefulness and power of numbers and extends their knowledge of the number system. It is critical in grades K-4 to develop concepts and relationships that will serve as a foundation for more advanced concepts and skills.

The K-4 instruction should help pupils understand fractions and decimals, explore their relationship, and build initial concepts about order and equivalence. Because evidence suggests that children construct these ideas slowly, it is crucial that teachers use physical materials, diagrams, and real-world situations in conjunction

Activities for Elementary School Mathematics

with ongoing efforts to relate their learning experiences to oral language and symbols. This K-4 emphasis on basic ideas will reduce the amount of time currently spent in the upper grades in correcting pupils' misconceptions and procedural difficulties.

STANDARD 13: PATTERNS AND RELATIONSHIPS
In grades K-4, the mathematics curriculum should include the study of patterns and relationships so that pupil can--
    recognize, describe, extend, and create a wide variety of patterns;
    represent and describe mathematical relationships;
    explore the use of variables and open sentences to express relationships.

Focus
    Patterns are everywhere. Children who are encouraged to look for patterns and to express them mathematically begin to understand how mathematics applies to the world in which they live. Identifying and working with a wide variety of patterns help children to develop the ability to classify and organize information. Relating patterns in numbers, geometry, and measurement helps them understand connections among mathematical topics. Such connections foster the kind of mathematical thinking that serves as a foundation for the more abstract ideas studied in later grades.
    From the earliest grades, the curriculum should give pupils opportunities to focus on regularities in events, shapes, designs, and sets of numbers. Children should begin to see that regularity is the essence of mathematics. The idea of a functional relationship can be intuitively developed through observations of regularity and work with generalizable patterns.
    Physical materials and pictorial displays should be used to help children recognize and create patterns and relationships. Observing varied representations of the same pattern helps children identify its properties. The use of letters and other symbols in generalizing descriptions of these properties prepares children to use variables in the future. This experience builds readiness for a generalized view of mathematics and the later study of algebra.

CURRICULUM STANDARDS FOR GRADES 5-8

Overview
Thirteen curriculum standards are presented for grades 5-8:
1. Mathematics as Problem Solving
2. Mathematics as Communication
3. Mathematics as Reasoning
4. Mathematical Connections

5. Number and Number Relationships
6. Number Systems and Number Theory
7. Computation and Estimation
8. Patterns and Functions
9. Algebra
10. Statistics
11. Probability
12. Geometry
13. Measurement

Mathematics is a useful, exciting, and creative area of study that can be appreciated and enjoyed by all pupils in grades 5-8. It helps them develop their ability to solve problems and reason logically. It offers to these curious, energetic pupils a way to explore and make sense of their world. However, many pupils view the current mathematics curriculum in grades 5-8 as irrelevant, dull, and routine. Instruction has emphasized computational facility at the expense of a broad, integrated view of mathematics and has reflected neither the vitality of the subject nor the characteristics of the pupils.

An ideal 5-8 mathematics curriculum would expand pupils' knowledge of numbers, computation, estimation, measurement, geometry, statistics, probability, patterns and functions, and the fundamental concepts of algebra. The need for this kind of broadened curriculum is acute. An examination of textbook series shows the repetition of topics, approach, and level of presentation in grade after grade. A comparison of the tables of contents shows little change over grades 5-8. It is even more disconcerting to realize that the very chapters that contain the most new material, such as probability, statistics, geometry, and prealgebra, are covered in the last half of the books--the sections most often skipped by teachers for lack of time. The result is an ineffective curriculum that rehashes material pupils already have seen. Such a curriculum promotes a negative image of mathematics and fails to give pupils an adequate background for secondary school mathematics.

These thirteen standards promote a broad curriculum for pupils in grades 5-8. Developing certain computational skills is important but constitutes only a part of this curriculum. Nevertheless, the existing curriculum in some schools prohibits many pupils from studying a broader curriculum until they have "mastered" basic computational skills. Shifting the focus to a broader curriculum is important for the following reasons:

1. Basic skills today and in the future mean far more than computational proficiency. Moreover, the calculator renders obsolete much of the complex paper-and-pencil proficiency traditionally emphasized in mathematics courses. Topics such as geometry, probability, statistics, and algebra have become increasingly more important and accessible to pupils through technology.

2. If pupils have not been successful in "mastering" basic computational skills in previous years, why should they be successful now, especially if the same methods that failed in the past are merely repeated? In fact, considering the effect of

failure on pupils' attitudes, we might argue that further efforts toward mastering computational skills are counterproductive.

3. Many of the mathematics topics that are omitted actually can help pupils recognize the need for arithmetic concepts and skills and provide fresh settings for their use. For example, in probability, pupils have many opportunities to add and multiply fractions.

The vision articulated in the 5-8 standards is of a broad, concept-driven curriculum, one that reflects the full breadth of relevant mathematics and its interrelationships with technology. This vision is built on five overall curricular goals for pupils: learning to value mathematics, becoming confident in their ability, becoming a mathematical problem solver, learning to communicate mathematically, and learning to reason mathematically. The teaching of this curriculum should be related to the characteristics of middle school pupils and their current and future needs.

Learner Characteristics

Implementation of the 5-8 standards should consider the unique characteristics of middle school pupils. As vast changes occur in their intellectual, psychological, social, and physical development, pupils in grades 5-8 begin to develop their abilities to think and reason more abstractly. Throughout this period, however, concrete experiences should continue to provide the means by which they construct knowledge. From these experiences they abstract more complex meanings and ideas. The use of language, both written and oral, helps pupils clarify their thinking and report their observations as they form and verify their mathematical ideas.

Pupils at this level can aptly be called "children in transition": they are restless, energetic, responsive to peer influence, and unsure about themselves. Self-consciousness is their hallmark, and curiosity about such questions as Who am I? How do I fit in? What do I enjoy doing? What do I want to be? is both their motivation and their nemesis. From this turmoil emerges an individual, with attitudes and patterns of thought taking shape.

In the transition to adulthood, middle school pupils are forming lifelong values and skills. The decisions pupils make about what they will study and how they will learn can dramatically affect their future. Failure to study mathematics can close the doors to vocational-technical schools, college majors, and careers--a loss of opportunity that happens most often to young women and minority pupils. Because many of the attitudes that affect these decisions are developed during the middle grades, it is crucial that conscious efforts be made to encourage all pupils, especially young women and minorities, to pursue mathematics. To this end, the curriculum must be interesting and relevant, must emphasize the usefulness of mathematics, and must foster a positive disposition toward mathematics. Whenever possible, pupils' cultural backgrounds should be integrated into the learning experience. Black or Hispanic pupils, for example, may find the development of mathematical ideas in their cultures of great interest. Teachers must also be sensitive to the fact that pupils bring very different everyday experiences to the mathematics classroom. The way in which a pupil from an urban environment and a pupil from a suburban or rural environment interpret a problem situation can be very different. This is an important reason why communication is one of the overarching goals of these standards.

Pupils will perform better and learn more in a caring environment in which they feel free to explore mathematical ideas, ask questions, discuss their ideas, and make mistakes. By listening to pupils' ideas and encouraging them to listen to one another, one can establish an atmosphere of mutual respect. Teachers can foster this willingness to share by helping pupils explore a variety of ideas in reaching solutions and verifying their own thinking. This approach instills in pupils an understanding of the value of independent learning and judgment and discourages them from relying on an outside authority to tell them whether they are right or wrong.

Features of the Mathematics Curriculum in Grades 5-8

Problem situations that establish the need for new ideas and motivate pupils should serve as the context for mathematics in grades 5-8. Although a specific idea might be forgotten, the context in which it is learned can be remembered and the idea re-created. In developing the problem situations, teachers should emphasize the application of mathematics to real-world problems as well as to other settings relevant to middle school pupils.

Communication with and about mathematics and mathematical reasoning should permeate the 5-8 curriculum. A broad range of topics should be taught, including number concepts, computation, estimation, functions, algebra, statistics, probability, geometry, and measurement. Although each of these areas is valid mathematics in its own right, they should be taught as an integrated whole, not as isolated topics; the connections among them should be a prominent feature of the curriculum.

Technology, including calculators, computers, and videos, should be used when appropriate. These devices and formats free pupils from tedious computations and allow them to concentrate on problem solving and other important content. They also give them new means to explore content. As paper-and-pencil computation becomes less important, the skills and understanding required to make proficient use of calculators and computers become more important.

Instruction

The standards are not intended to each constitute a chapter in a text or a particular unit of instruction; rather, learning activities should incorporate topics and ideas across standards. For example, an instructional activity might involve problem solving and use geometry, measurement, and computation. All mathematics should be studied in contexts that give the ideas and concepts meaning. Problems should arise from situations that are not always well formed. pupils should have opportunities to formulate problems and questions that stem from their own interests.

Learning should engage pupils both intellectually and physically. They must become active learners, challenged to apply their prior knowledge and experience in new and increasingly more difficult situations. Instructional approaches should engage pupils in the process of learning rather than transmit information for them to receive. Middle grade pupils are especially responsive to hands-on activities in tactile, auditory, and visual instructional modes. Classroom activities should provide pupils the opportunity to work both individually and in small- and large-group arrangements. The arrangement should be determined by the instructional goals as well as the nature of the activity. Individual work can help pupils develop confidence in their own ability to solve problems but should constitute only a portion of the middle school experience. Working in small groups provides pupils with opportunities to talk about ideas and listen to their peers, enables teachers to interact more closely with pupils, takes positive advantage of the social characteristics of the middle school pupil, and provides opportunities for pupils to exchange ideas and hence develops their ability to communicate and reason. Small-group work can involve collaborative or cooperative as well as independent work. Projects and small-group work can empower pupils to become more independent in their own learning. Whole-class discussions require pupils to synthesize, critique, and summarize strategies, ideas, or conjectures that are the products of individual and group work. These mathematical ideas can be expanded to, and integrated with, other subjects.

Materials

The 5-8 standards make the following assumptions about classroom materials:

Every classroom will be equipped with ample sets of manipulative materials and supplies (e.g., spinners, cubes, tiles, geoboards, pattern blocks, scales, compasses, scissors, rulers, protractors, graph paper, grid-and-dot paper).

Teachers and pupils will have access to appropriate resource materials from which to develop problems and ideas for explorations.

All pupils will have a calculator with functions consistent with the tasks envisioned in this curriculum. Calculators should include the following features: algebraic logic including order of operations; computation in decimal and common fraction form; constant function for addition, subtraction, multiplication, and division; and memory, percent, square root, exponent, reciprocal, and +/- keys.

Every classroom will have at least one computer available at all times for demonstrations and pupil use. Additional computers should be available for individual, small-group, and whole-class use.

SUMMARY OF CHANGES IN CONTENT AND EMPHASIS IN 5--8 MATHEMATICS

GIVE MORE ATTENTION TO:

PROBLEM SOLVING
- Pursuing open-ended problems and extended problem-solving projects
- Investigating and formulating questions from problem situations
- Representing situations verbally, numerically, graphically, geometrically, or symbolically

COMMUNICATION
- Discussing, writing, reading, and listening to mathematical ideas

REASONING
- Reasoning in spatial contexts
- Reasoning with proportions
- Reasoning from graphs
- Reasoning inductively and deductively

CONNECTIONS
- Connecting mathematics to other subjects and to the world outside the classroom
- Connecting topics within mathematics
- Applying mathematics

NUMBER/OPERATIONS/COMPUTATION
- Developing number sense
- Developing operation sense
- Creating algorithms and procedures
- Using estimation both in solving problems and in checking the reasonableness of results
- Exploring relationships among representations of, and operations on, whole numbers, fractions, decimals, integers, and rational numbers
- Developing an understanding of ratio, proportion, and percent

PATTERNS AND FUNCTIONS
- Identifying and using functional relationships
- Developing and using tables, graphs, and rules to describe situations
- Interpreting among different mathematical representations

ALGEBRA
- Developing an understanding of variables, expressions, and equations
- Using a variety of methods to solve linear equations and informally investigate inequalities and nonlinear equations

STATISTICS
- Using statistical methods to describe, analyze, evaluate, and make decisions

PROBABILITY
> Creating experimental and theoretical models of situations involving probabilities

GEOMETRY
> Developing an understanding of geometric objects and relationships
> Using geometry in solving problems

MEASUREMENT
> Estimating and using measurement to solve problems

INSTRUCTIONAL PRACTICES
> Actively involving pupils individually and in groups in exploring, conjecturing, analyzing, and applying mathematics in both a mathematical and a real-world context
> Using appropriate technology for computation and exploration
> Using concrete materials
> Being a facilitator of learning
> Assessing learning as an integral part of instruction

GIVE LESS ATTENTION TO:

PROBLEM SOLVING
> Practicing routine, one-step problems
> Practicing problems categorized by types (e.g., coin problems, age problems)

COMMUNICATION
> Doing fill-in-the-blank worksheets
> Answering questions that require only yes, no, or a number as responses

REASONING
> Relying on outside authority (teacher or an answer key)

CONNECTIONS
> Learning isolated topics
> Developing skills out of context

NUMBER/OPERATIONS/COMPUTATION
> Memorizing rules and algorithms
> Practicing tedious paper-and-pencil computations
> Finding exact forms of answers
> Memorizing procedures, such as cross-multiplication, without understanding
> Practicing rounding numbers out of context

PATTERNS AND FUNCTIONS
> Topics seldom in the current curriculum

ALGEBRA
> Manipulating symbols
> Memorizing procedures and drilling on equation solving

STATISTICS
    Memorizing formulas
PROBABILITY
    Memorizing formulas
GEOMETRY
    Memorizing geometric vocabulary
    Memorizing facts and relationships
MEASUREMENT
    Memorizing and manipulating formulas
    Converting within and between measurement systems
INSTRUCTIONAL PRACTICES
    Teaching computations out of context
    Drilling on paper-and-pencil algorithms
    Teaching topics in isolation
    Stressing memorization
    Being the dispenser of knowledge
    Testing for the sole purpose of assigning grades

STANDARD 1: MATHEMATICS AS PROBLEM SOLVING

In grades 5-8, the mathematics curriculum should include numerous and varied experiences with problem solving as a method of inquiry and application so that pupils can--
    use problem-solving approaches to investigate and understand mathematical content;
    formulate problems from situations within and outside mathematics;
    develop and apply a variety of strategies to solve problems, with emphasis on multistep and nonroutine problems;
    verify and interpret results with respect to the original problem situation;
    generalize solutions and strategies to new problem situations;
    acquire confidence in using mathematics meaningfully.

Focus

    Problem solving is the process by which pupils experience the power and usefulness of mathematics in the world around them. It is also a method of inquiry and application, interwoven throughout the Standards to provide a consistent context for learning and applying mathematics. Problem situations can establish a "need to know" and foster the motivation for the development of concepts.
    In grades 5-8, the curriculum should take advantage of the expanding mathematical capabilities of middle school pupils to include more complex problem situations involving topics such as probability, statistics, geometry, and rational numbers. Situations and approaches should build on and extend the mathematical

language pupils are acquiring and help them to develop a variety of problem-solving strategies and approaches. Although concrete and empirical situations remain a focus throughout these grades, a balance should be struck between problems that apply mathematics to the real world and problems that arise from the investigation of mathematical ideas. Finally, the mathematics curriculum should engage pupils in some problems that demand extended effort to solve. Some might be group projects that require pupils to use available technology and to engage in cooperative problem solving and discussion. For grades 5-8 an important criterion of problems is that they be interesting to pupils. Computers and calculators are powerful problem-solving tools. The power to compute rapidly, to graph a relationship instantly, and to systematically change one variable and observe what happens to other related variables can help pupils become independent doers of mathematics.

STANDARD 2: MATHEMATICS AS COMMUNICATION

In grades 5-8, the study of mathematics should include opportunities to communicate so that pupils can--
- model situations using oral, written, concrete, pictorial, graphical, and algebraic methods;
- reflect on and clarify their own thinking about mathematical ideas and situations;
- develop common understandings of mathematical ideas, including the role of definitions;
- use the skills of reading, listening, and viewing to interpret and evaluate mathematical ideas;
- discuss mathematical ideas and make conjectures and convincing arguments;
- appreciate the value of mathematical notation and its role in the development of mathematical ideas.

Focus

The use of mathematics in other disciplines has increased dramatically, largely because of its power to represent and communicate ideas concisely. Society's increasing use of technology requires that pupils learn both to communicate with computers and to make use of their own individual power as a medium of communication. The ability to read, write, listen, think creatively, and communicate about problems will develop and deepen pupils' understanding of mathematics.

Middle school pupils should have many opportunities to use language to communicate their mathematical ideas. The communication process requires pupils to reach agreement about the meanings of words and to recognize the crucial importance of commonly shared definitions. Opportunities to explain, conjecture, and defend one's ideas orally and in writing can stimulate deeper understandings of concepts and principles. It is essential that mathematical concepts be firmly attached to the symbols that represent them; the need for symbolic representation arises out of the exploration of these concepts. In the process of discussing mathematical concepts and symbols, pupils become aware of the connections between them. Unless pupils frequently and explicitly discuss relationships between concepts and symbols, they are likely to view symbols as disparate objects to be memorized.

As pupils progress from grade 5 to grade 8, their ability to reason abstractly matures greatly. Concurrent with this enhanced ability to abstract common elements from situations, to conjecture, and to generalize--in short, to do mathematics--should come an increasing sophistication in the ability to communicate mathematics. But this development cannot occur without deliberate and careful acquisition of the language of mathematics.

STANDARD 3: MATHEMATICS AS REASONING
In grades 5-8, reasoning shall permeate the mathematics curriculum so that pupils can--
- recognize and apply deductive and inductive reasoning;
- understand and apply reasoning processes, with special attention to spatial reasoning and reasoning with proportions and graphs;
- make and evaluate mathematical conjectures and arguments;
- validate their own thinking;
- appreciate the pervasive use and power of reasoning as a part of mathematics.

Focus

Reasoning is fundamental to the knowing and doing of mathematics. Although most disciplines have standards of evaluation by which new theories or discoveries are judged, nowhere are these standards as explicit and well formulated as they are in mathematics. Conjecturing and demonstrating the logical validity of conjectures are the essence of the creative act of doing mathematics. To give more pupils access to mathematics as a powerful way of making sense of the world, it is essential that an emphasis on reasoning pervade all mathematical activity. Pupils need a great deal of time and many experiences to develop their ability to construct valid arguments in problem settings and evaluate the arguments of others.

The development of logical reasoning is tied to the intellectual and verbal development of pupils. Through grades 5-8, pupils' reasoning abilities change. Whereas most fifth graders still are concrete thinkers who depend on a physical or concrete context for perceiving regularities and relationships, many eighth-grade pupils are capable of more formal reasoning and abstraction. Even the most advanced pupils at the 5-8 level, however, might use concrete materials to support their reasoning; this is especially true for spatial reasoning. The 5-8 mathematics curriculum should pay special attention to the development of pupil's abilities to use proportional and spatial reasoning and to reason from graphs.

Technology can foster environments in which pupils' growing curiosity can lead to rich mathematical invention. In these environments, the control of exploring mathematical ideas is turned over to pupils. Both inductive and deductive reasoning come into play as pupils make conjectures and seek to explain why they are valid. Whether encouraged by technology or by challenging mathematical situations posed in the classroom, this freedom to explore, conjecture, validate, and to convince others is critical to the development of mathematical reasoning in the middle grades.

STANDARD 4: MATHEMATICAL CONNECTIONS
In grades 5-8, the mathematics curriculum should include the investigation of mathematical connections so that pupils can--
- see mathematics as an integrated whole;
- explore problems and describe results using graphical, numerical, physical, algebraic, and verbal mathematical models or representations;
- use a mathematical idea to further their understanding of other mathematical ideas;
- apply mathematical thinking and modeling to solve problems that arise in other disciplines, such as art, music, psychology, science, and business;
- value the role of mathematics in our culture and society.

Focus

For many pupils, mathematics in the middle grades has far too often simply repeated or extended much of the computational work covered in the earlier grades. The intent of this standard is to help pupils broaden their perspective, to view mathematics as an integrated whole rather than as an isolated set of topics, and to acknowledge its relevance and usefulness both in and out of school. Mathematics instruction at the 5-8 level should prepare pupils for expanded and deeper study in high school through exploration of the interconnections among mathematical ideas.

Pupils should have many opportunities to observe the interaction of mathematics with other school subjects and with everyday society. To accomplish this, mathematics teachers must seek and gain the active participation of teachers of other disciplines in exploring mathematical ideas through problems that arise in their classes. This integration of mathematics into contexts that give its symbols and processes practical meaning is an overarching goal of all the standards. It allows pupils to see how one mathematical idea can help them understand others, and it illustrates the subject's usefulness in solving problems, describing and modeling real-world phenomena, and communicating complex thoughts and information in a concise and precise manner. Different representations of problems serve as different lenses through which pupils interpret the problems and the solutions. If pupils are to become mathematically powerful, they must be flexible enough to approach situations in a variety of ways and recognize the relationships among different points of view.

STANDARD 5: NUMBER AND NUMBER RELATIONSHIPS
In grades 5-8, the mathematics curriculum should include the continued development of number and number relationships so that pupils can--
> understand, represent, and use numbers in a variety of equivalent forms (integer, fraction, decimal, percent, exponential, and scientific notation) in real-world and mathematical problem situations;
> develop number sense for whole numbers, fractions, decimals, integers, and rational numbers;
> understand and apply ratios, proportions, and percents in a wide variety of situations;
> investigate relationships among fractions, decimals, and percents;
> represent numerical relationships in one- and two-dimensional graphs.

Focus

The use of concise symbols and language to represent numbers is a significant historical and practical development. In the middle school years, pupils come to recognize that numbers have multiple representations, so the development of concepts for fractions, ratios, decimals, and percents and the idea of multiple representations of these numbers need special attention and emphasis. The ability to generate, read, use, and appreciate multiple representations of the same quantity is a critical step in learning to understand and do mathematics.

As pupils progress through middle school, they build their knowledge of rational numbers as important both in their own right and as a foundation for rational forms in algebra. Ratio, proportion, and percent are introduced and developed in grades 5-8. In exploring these topics, pupils have many opportunities to develop their ability to reason proportionally.

STANDARD 6: NUMBER SYSTEMS AND NUMBER THEORY
In grades 5-8, the mathematics curriculum should include the study of number systems and number theory so that pupils can--
> understand and appreciate the need for numbers beyond the whole numbers;
> develop and use order relations for whole numbers, fractions, decimals, integers, and rational numbers;
> extend their understanding of whole number operations to fractions, decimals, integers, and rational numbers;
> understand how the basic arithmetic operations are related to one another;
> develop and apply number theory concepts (e.g., primes, factors, and multiples) in real-world and mathematical problem situations.

Focus

The central theme of this standard is the underlying structure of mathematics, which bonds its many individual facets into a useful, interesting, and logical whole. Instruction in grades 5-8 typically devotes a great deal of time to helping pupils master myriad details but pays scant attention to how these individual facets fit together. It is the intent of this standard that pupils should come to understand and appreciate mathematics as a coherent body of knowledge rather than a vast, perhaps bewildering, collection of isolated facts and rules. Understanding this structure promotes pupils' efficiency in investigating the arithmetic of fractions, decimals, integers, and rationals through the unity of common ideas. It also offers insights into how the whole number system is extended to the rational number system and beyond. It improves problem-solving capability by providing a better perspective of arithmetic operations.

Instruction that facilitates pupils' understanding of the underlying structure of arithmetic should employ informal explorations and emphasize the reasons why various kinds of numbers occur, commonalities among various arithmetic processes, and relationships between number systems.

STANDARD 7: COMPUTATION AND ESTIMATION

In grades 5-8, the mathematics curriculum should develop the concepts underlying computation and estimation in various contexts so that pupils can--

    compute with whole numbers, fractions, decimals, integers, and rational numbers;

    develop, analyze, and explain procedures for computation and techniques for estimation;

    develop, analyze, and explain methods for solving proportions;

    select and use an appropriate method for computing from among mental arithmetic, paper-and-pencil, calculator, and computer methods;

    use computation, estimation, and proportions to solve problems;

    use estimation to check the reasonableness of results.

Focus

Although computation is vital in this information age, technology has drastically changed the methods by which we compute. Whereas inexpensive calculators execute routine computations accurately and quickly and computers execute more complex computations with ease, many current mathematics programs focus on traditional paper-and-pencil algorithms. This standard prepares pupils to select and use appropriate mental, paper-and-pencil, calculator, and computer methods. It is no longer necessary or useful to devote large portions of instructional time to performing routine computations by hand. Other mathematical experiences for middle school pupils deserve far more emphasis. The facility in computation that calculators and computers offer should imbue the curriculum with an expanded potential for interesting problem-solving experiences. Pupils need more experience in developing procedures and evaluating their work and in interpreting the results of computations done by machines.

STANDARD 8: PATTERNS AND FUNCTIONS

In grades 5-8, the mathematics curriculum should include explorations of patterns and functions so that pupils can--

> describe, extend, analyze, and create a wide variety of patterns;
> describe and represent relationships with tables, graphs, and rules;
> analyze functional relationships to explain how a change in one quantity results in a change in another;
> use patterns and functions to represent and solve problems.

Focus

One of the central themes of mathematics is the study of patterns and functions. This study requires pupils to recognize, describe, and generalize patterns and build mathematical models to predict the behavior of real-world phenomena that exhibit the observed pattern. The widespread occurrence of regular and chaotic pattern behavior makes the study of patterns and functions important. Exploring patterns helps pupils develop mathematical power and instills in them an appreciation for the beauty of mathematics.

The study of patterns in grades 5-8 builds on pupils' experiences in K-4 but shifts emphasis to an exploration of functions. However, work with patterns continues to be informal and relatively unburdened by symbolism. Pupils have opportunities to generalize and describe patterns and functions in many ways and to explore the relationships among them. When pupils make graphs, data tables, expressions, equations, or verbal descriptions to represent a single relationship, they discover that different representations yield different interpretations of a situation.

In informal ways, pupils develop an understanding that functions are composed of variables that have a dynamic relationship: Changes in one variable result in change in another. The identification of the special characteristics of a relationship, such as minimum or maximum values or points at which the value of one of the variables is 0 (x- and y-intercepts), lays the foundation for a more formal study of functions in grades 9-12.

The theme of patterns and functions is woven throughout the 5-8 standards. It begins in K-4, is extended and made more central in 5-8, and reaches maturity with a natural extension to symbolic representation and supporting concepts, such as domain and range, in grades 9-12. Examples appropriate for grades 5-8 are incorporated into other standards for this age group.

STANDARD 9: ALGEBRA

In grades 5-8, the mathematics curriculum should include explorations of algebraic concepts and processes so that pupils can-
- understand the concepts of variable, expression, and equation;
- represent situations and number patterns with tables, graphs, verbal rules, and equations and explore the interrelationships of these representations;
- analyze tables and graphs to identify properties and relationships;
- develop confidence in solving linear equations using concrete, informal, and formal methods;
- investigate inequalities and nonlinear equations informally;
- apply algebraic methods to solve a variety of real-world and mathematical problems.

Focus

The middle school mathematics curriculum is, in many ways, a bridge between the concrete elementary school curriculum and the more formal mathematics curriculum of the high school. One critical transition is that between arithmetic and algebra. It is thus essential that in grades 5-8, pupils explore algebraic concepts in an informal way to build a foundation for the subsequent formal study of algebra. Such informal explorations should emphasize physical models, data, graphs, and other mathematical representations rather than facility with formal algebraic manipulation. Pupils should be taught to generalize number patterns to model, represent, or describe observed physical patterns, regularities, and problems. These informal explorations of algebraic concepts should help pupils to gain confidence in their ability to abstract relationships from contextual information and use a variety of representations to describe those relationships.

Activities in grades 5-8 should build on pupils' K-4 experiences with patterns. They should continue to emphasize concrete situations that allow pupils to investigate patterns in number sequences, make predictions, and formulate verbal rules to describe patterns. Learning to recognize patterns and regularities in mathematics and make generalizations about them requires practice and experience. Expanding the amount of time that pupils have to make this transition to more abstract ways of thinking increases their chances of success. By integrating informal algebraic experiences throughout the K-8 curriculum, pupils will develop confidence in using algebra to represent and solve problems. In addition, by the end of the eighth grade, pupils should be able to solve linear equations by formal methods and some nonlinear equations by informal means.

STANDARD 10: STATISTICS

In grades 5-8, the mathematics curriculum should include exploration of statistics in real-world situations so that pupils can--
- systematically collect, organize, and describe data;

construct, read, and interpret tables, charts, and graphs;
make inferences and convincing arguments that are based on data analysis;
evaluate arguments that are based on data analysis;
develop an appreciation for statistical methods as powerful means for decision making.

Focus

In this age of information and technology, an ever-increasing need exists to understand how information is processed and translated into usable knowledge. Because of society's expanding use of data for prediction and decision making, it is important that pupils develop an understanding of the concepts and processes used in analyzing data. A knowledge of statistics is necessary if pupils are to become intelligent consumers who can make critical and informed decisions.

In grades K-4, pupils begin to explore basic ideas of statistics by gathering data appropriate to their grade level, organizing them in charts or graphs, and reading information from displays of data. These concepts should be expanded in the middle grades. Pupils in grades 5-8 have a keen interest in trends in music, movies, fashion, and sports. An investigation of how such trends are developed and communicated is an excellent motivator for the study of statistics. Pupils need to be actively involved in each of the steps that comprise statistics, from gathering information to communicating results.

STANDARD 11: PROBABILITY

In grades 5-8, the mathematics curriculum should include explorations of probability in real-world situations so that pupils can--
model situations by devising and carrying out experiments or simulations to determine probabilities;
model situations by constructing a sample space to determine probabilities;
appreciate the power of using a probability model by comparing experimental results with mathematical expectations;
make predictions that are based on experimental or theoretical probabilities;
develop an appreciation for the pervasive use of probability in the real world.

Focus

Probability theory is the buttress of the modern world. Current research in both the physical and social sciences cannot be understood without it. Today's politics, tomorrow's weather report, and next week's satellites all depend on it.

An understanding of probability and the related area of statistics is essential to being an informed citizen. Often we read statements such as, "There is a 20 percent chance of rain or snow today." "The odds are three to two that the Cats will win the championship." "The probability of winning the grand prize in the state lottery is 1 in 7, 240, 000." Pupils in the middle grades have an intense interest in the notions

Activities for Elementary School Mathematics

of fairness and the chances of winning games. The study of probability develops concepts and methods for investigating such situations. These methods allow pupils to make predictions when uncertainty exists and to make sense of claims that they see and hear.

The study of probability in grades 5-8 should not focus on developing formulas or computing the likelihood of events pictured in texts. Pupils should actively explore situations by experimenting and simulating probability models. Such investigations should embody a variety of realistic problems, from questions about sports events to whether it will rain on the day of the school picnic. Pupils should talk about their ideas and use the results of their experiments to model situations or predict events. Probability is rich in interesting problems that can fascinate pupils and provide settings for developing or applying such concepts as ratios, fractions, percents, and decimals.

STANDARD 12: GEOMETRY

In grades 5-8, the mathematics curriculum should include the study of the geometry of one, two, and three dimensions in a variety of situations so that pupils can--
- identify, describe, compare, and classify geometric figures;
- visualize and represent geometric figures with special attention to developing spatial sense;
- explore transformations of geometric figures;
- represent and solve problems using geometric models;
- understand and apply geometric properties and relationships;
- develop an appreciation of geometry as a means of describing the physical world.

Focus

Geometry is grasping space . . . that space in which the child lives, breathes and moves. The space that the child must learn to know, explore, conquer, in order to live, breathe and move better in it. The study of geometry helps pupils represent and make sense of the world. Geometric models provide a perspective from which pupils can analyze and solve problems, and geometric interpretations can help make an abstract (symbolic) representation more easily understood. Many ideas about number and measurement arise from attempts to quantify real-world objects that can be viewed geometrically. For example, the use of area models provides an interpretation for much of the arithmetic of decimals, fractions, ratios, proportions, and percents.

Pupils discover relationships and develop spatial sense by constructing, drawing, measuring, visualizing, comparing, transforming, and classifying geometric figures. Discussing ideas, conjecturing, and testing hypotheses precede the development of more formal summary statements. In the process, definitions become meaningful, relationships among figures are understood, and pupils are prepared to use these ideas to develop informal arguments. The informal exploration of geometry can be exciting and mathematically productive for middle school pupils. At this level, geometry should focus on investigating and using geometric ideas and relationships rather than on memorizing definitions and formulas. The study of geometry in grades 5-8 links the informal explorations begun in grades K-4 to the more formalized processes studied in grades 9-12. The expanding logical capabilities of pupils in grades 5-8 allow them to draw inferences and make logical deductions from geometric problem situations. This does not imply that the study of geometry in grades 5-8 should be a formalized endeavor; rather, it should simply provide increased opportunities for pupils to engage in more systematic explorations.

STANDARD 13: MEASUREMENT

In grades 5-8, the mathematics curriculum should include extensive concrete experiences using measurement so that pupils can--

- extend their understanding of the process of measurement;
- estimate, make, and use measurements to describe and compare phenomena;
- select appropriate units and tools to measure to the degree of accuracy required in a particular situation;
- understand the structure and use of systems of measurement;
- extend their understanding of the concepts of perimeter, area, volume, angle measure, capacity, and weight and mass;
- develop the concepts of rates and other derived and indirect measurements;
- develop formulas and procedures for determining measures to solve problems.

Focus

Measurement activities can and should require a dynamic interaction between pupils and their environment. Pupils encounter measurement ideas both in and out of school, in such areas as architecture, art, science, commercial design, sports, cooking, shopping, and map reading. The study of measurement shows the usefulness and practical applications of mathematics, and pupils' need to communicate about various measurements highlights the importance of standard units and common measurement systems.

Measurement in grades 5-8 should be an active exploration of the real world. As pupils acquire the ability to use appropriate tools in measuring objects, they should extend these skills to new situations and new applications. The approximate nature of measure is an aspect of number that deserves repeated attention at this level. However, measurement activities in these grades should focus on using concepts and skills to solve problems and investigate other mathematical situations.

The development of the concepts of perimeter, area, volume, angle measure, capacity, and weight is initiated in grades K-4 and extended and applied in grades 5-8. At this level, pupils can begin to estimate the error of a measurement, adding to the

K-4 notion of "about" 4 cm. From their explorations, pupils should develop multiplicative procedures and formulas for determining measures. The curriculum should focus on the development of understanding, not on the rote memorization of formulas. In addition, the concepts of rate as a measure and of indirect measurement are developed in grades 5-8.

Geometry and measurement are interconnected and support each other in many ways. The concept of similarity, for example, can be used in indirect measurement, and the perimeter and area of irregular figures can be determined using line segments and squares, respectively. Measurement also has strong connections to the pupils' expanding concept of number. Fractions, decimals, and rational numbers are used to represent measures.

Using the Standards

The National Council of Teachers of Mathematics' Curriculum and Evaluation Standards for School Mathematics reflects the importance of understanding the development of knowledge at the K-4 level. The standards document recognizes that current instructional and curricular content must focus on students' active construction of mathematical knowledge. Instructional practices need to be conceptually oriented, involve children actively, emphasize the development of mathematical thinking and application, and include a broad range of content.[2]

When implementing a thematic approach, teachers need to furnish guided experiences related to the chosen theme. The NCTM's curriculum standards furnish a framework for organizing the mathematics content within a thematic unit. A thematic approach offers techniques for applying mathematical content that can engage the children's minds. After the selection of a theme or topic (e.g.; apples, trains, cars, bears, etc.), the teacher and students can design exploratory situations that allow students to engage in interactive experiences. These experiences should ideally evolve naturally and stimulate the development of mathematical thinking in the thirteen areas outlined in the K-4 standards. By using the planning process outlined here, a teacher can be sure that course content includes these important curricular goals for mathematics. The thematic approach is a method of instruction teachers use to allow children to engage actively in activities that focus on a topic that children, preferably, have selected to study.[3]

---

[2] Smith, Stephanie Z. et al., "What the NCTM Standards Look Like in One Classroom," *Educational Leadership* 50, May (1993): 4-7.
[3] Piazza, Jenny A. et al., "Thematic Webbing and the Curriculum Standards in the Primary Grades," *Arithmetic Teacher* 41, no. 6 (1994): 294-298.

Effective Math Teaching

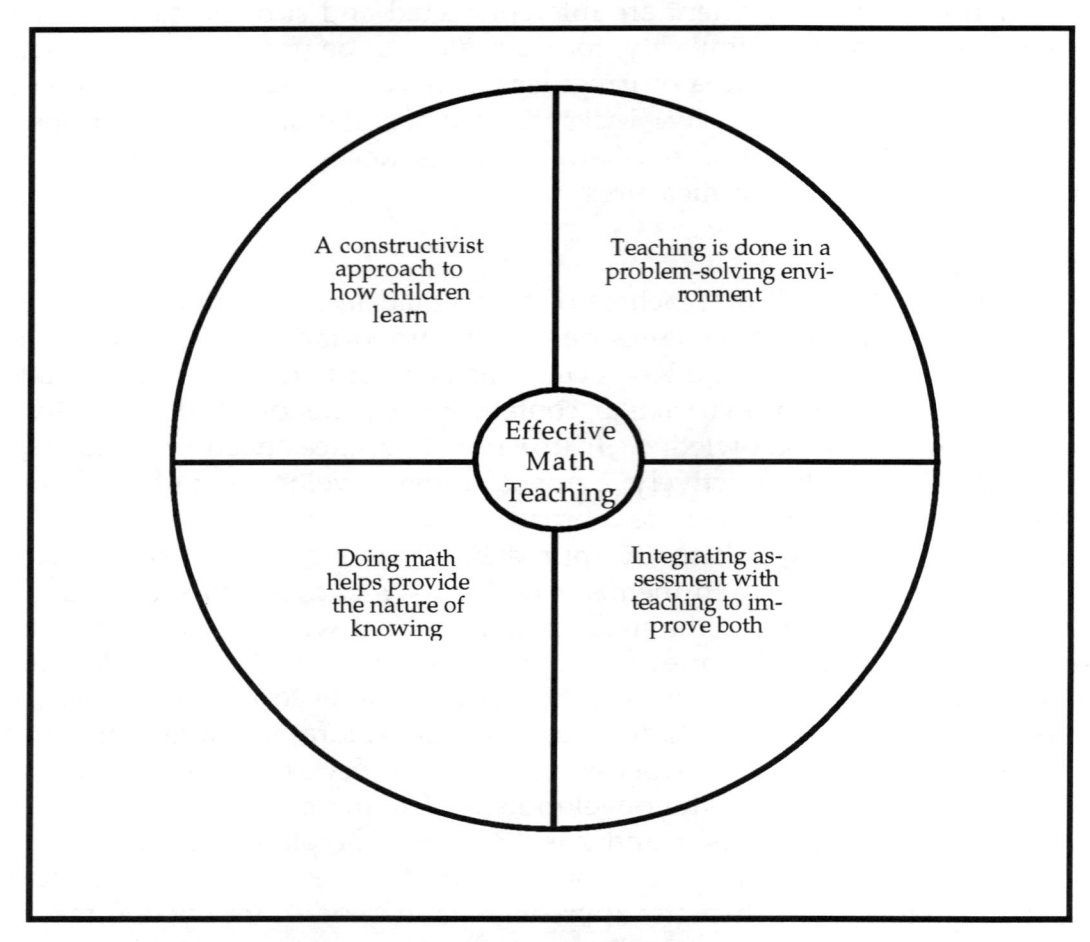

- A constructivist approach to how children learn
- Teaching is done in a problem-solving environment
- Doing math helps provide the nature of knowing
- Integrating assessment with teaching to improve both

NCTM Standards Provide Direction for Reforming Mathematics Teaching

THE NATIONAL COUNCIL OF TEACHERS OF MATHEMATICS

The National Council of Teachers of Mathematics (NCTM) is a nonprofit professional association founded in 1920 and is the largest mathematics organization in the world. NCTM is designed primarily for teachers who teach mathematics in grades K-14, and it has a membership of approximately 120,000 teachers, educators, and other professionals. Its purpose is to improve mathematics education for all students in the United States and Canada.

The primary purpose of NCTM is to provide leadership in the improvement of the teaching and learning of mathematics. To stimulate students' interest and accomplishments in mathematics and to promote a comprehensive education for every child, the Council has established three goals:

1. To foster excellence in school mathematics curricula and instructional programs, including assessment and evaluation
2. To promote professional excellence in mathematics teaching
3. To strengthen NCTM's leadership in mathematics education

The National Council of Teachers of Mathematics is headquartered at 1906 Association Drive, Reston, Virginia 20191-1593. Phone: (703) 620-9840; Fax: (703) 476-2970; Toll Free: (800) 235-7566. The internet site for NCTM is: http://www.nctm.org/index.htm, and their e-mail address is nctm@nctm.org.

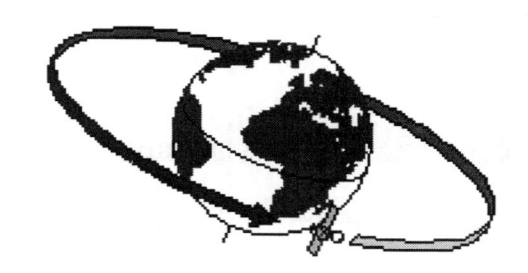

MATHEMATICS RESOURCES ON THE INTERNET

http://www.nctm.org/index.htm
   The National Council of Teachers of Mathematics homepage, containing information about the NCTM and its publications, events, and math products.

http://www.planetary.caltech.edu/~eww/math/math.html
   Eric's Treasure Trove of Mathematics. Anything you ever wanted to know about any math topic but were afraid to ask.

http://www.forum.swarthmore.edu/
   A resource for anyone interested in mathematics education.

http://www.tc.cornell.edu/Edu/MathSciGateway/math.html
   A great connection to other math sites.

http://archives.math.utk.edu/k12.html
   Mathematics Archives–K-12 Internet Sites. A collection of materials to be used in teaching math.

http://www.edfac.unimelb.edu.au/DSME/research/TAME/TAME_LINKS.html
   This page contains links to internet-based resources that may have some use for mathematics education.

http://explorer.scrtec.org/explorer
   Explorer Home Page. A collection of educational resources including lesson plans, instructional software, and lab activities for K-12 math and science education.

http://204.161.33.100/aims.html/
   A meeting place for math educators to exchange ideas about teaching, lessons, activities, and teaching products.

http://www.mste.uiuc.edu/mathed/queryform.html#search
   Mathematics Lessons Database. A search tool for internet-based math lessons and activities.

http://www.nptn.org/cyber.serv/AOneP/academy_one/teacher/cec/cecmath/math-elem.html
   Mini-lessons in math appropriate for grades K-8.

http://www.cs.rice.edu/~sboone/Lessons/lptitle.html
   Lessons developed under a National Science Foundation program called Girl TECH '95, a teacher training and student technology council program.

http://www.pacificnet.net/~mandel/index.html
   Teachers Helping Teachers. This site lists lessons and teaching tips submitted by users and updated weekly.

http://acorn.educ.nottingham.ac.uk/Maths/photomath/welcome.html
   Math with Photographs. Mathematical problems are presented pertaining to given pictures.

http://www.elk-grove.k12.il.us/schoolweb/highland/highland.tess.html
   Math Tessellations. This unit was developed in order for students to realize the importance of mathematics in the art world. Tessellations are defined and investigated by students with the use of attribute blocks.

http://www-sci.lib.uci.edu/SEP/math.html
   Frank Potter's Science Gems–Mathematics. This site has math resources for grades K-16.

http://www.nova.edu/Inter-Links/puzzles.html
   This site presents puzzles and brain teasers which are fun for all ages.

http://www.hmco.com:80/school/math/brain/
   This site posts puzzles for ages 3-4, 5-6, and 7+. New puzzles appear weekly.

http://www.saxonpub.com/text_graphics/Math_Challenge.html
   This site provides daily math puzzles and challenges.

http://www.sisweb.com/math/tables.htm
   This site has a large list of famous and commonly used formulas and tables.

http://www.csun.edu/~hcmth014/comics.html
   This site contains a large collection of math jokes.

Activities for Elementary School Mathematics

http://forum.swarthmore.edu/dr.math/dr-math.html
   This site is a list of questions submitted by users and Dr. Math's answers.

http://math.furman.edu/~mwoodard/mquot.html
   This site provides mathematics quotations arranged alphabetically by author.

http://archives.math.utk.edu/k12_software.html
   This site lists math-related software packages and their sources.

http://www.mathpro.com/math/glossary/glossary.html
   On-line Mathematics Dictionary. This site lists terms alphabetically with their definitions.

http://acorn.educ.nottingham.ac.uk/cgi-bin/daynum
   Today's Date. Interesting facts about today's date are given at this site.

http://www-groups.dcs.st-and.ac.uk:80/~history/
   History of Mathematics. This site gives the history behind any math topic, from Aristotle to Zeno.

http://www.scottlan.edu/lriddle/women/women.html
   Women Mathematicians. This site gives biographies, including pictures, of women mathematicians.

Constructivism

In today's educational community, a theory known as *constructivism* is perhaps the most widely accepted view of how children learn. It proposes that children must be active participants in the development of their own understanding.

Deeply rooted in the cognitive theories of Piaget,[1] which date from the 1960s, constructivism rejects the notion that children's minds are blank slates awaiting something to be imprinted. Instead, constructivism suggests that children help create their own knowledge through the basic mental activities of *assimilation* and *accommodation*.

Stemming from the view that the mind constantly changes its structure to help us make sense of things that we perceive, Piaget described the shifting modifications in the mind as assimilation and accommodation. When things are familiar to us and fit well with what we already know from our previously developed understanding, we fit the new experiences into our existing ideas. Piaget called this process *assimilation*. When new experiences or perceptions do not fit with what we already know, however, Piaget said that we must appease the dissonance within our minds by modifying the new experience to make it fit. Piaget called this modifying process *accommodation*, a shift in our cognitive framework which permits assimilation of the new experience. Knowledge is built up by the learner as existing ideas are expanded, elaborated, and changed to allow a new idea to fit.

In the constructivist approach, then, learners actively build their own understanding through reflective thought. Reflective thinking is stimulated as learners brainstorm in an environment of encouragement (provided by the teacher), exploring the networks of ideas already existing in their minds. They integrate these networks, both within their own minds and by sharing with other learners through active reflective thought. The networks of ideas change, become rearranged, take on additions, and are modified as learning occurs. The more connections the learner can make with his/her existing network of ideas, the better that new experiences and ideas are understood (or learned).

Constructivist instruction is based on the supposition that the pupil is a naturally active learner who constructs new individual knowledge through linking prior knowledge with new knowledge. Constructivist learning involves an interactive and collaborative dialogue between the teacher, the pupil, and other learners. The teacher orchestrates the learning by providing a rich and supportive environment where assistance and direction allow the learner to construct his/her own knowledge. This construction of knowledge results in ownership by the learner and, thus, a deeper understanding of the new information. The teacher focuses on guiding the learner to achieve success. The learner is a proactive participant in the learning process and not a sponge waiting to absorb knowledge!

---

[1] Piaget, J. (1977). *The development of thought: Equilibrium of cognitive structures.* New York: Viking Press.

Multiple Intelligences

Howard Gardner[1] theorized that humans have multiple intelligences, and he identified seven: bodily-kinesthetic intelligence, interpersonal intelligence, intrapersonal intelligence, linguistic intelligence, logical-mathematical intelligence, musical intelligence, and spatial intelligence.

Bodily-kinesthetic intelligence, said Gardner, relates to physical movement and the ability to use and master one's bodily motions in highly skilled ways. Gardner identified **interpersonal intelligence** as ability and skill in person-to-person relationships such as understanding and dealing effectively with another person's moods, temperaments, behaviors, feelings, and motivations. **Intrapersonal intelligence** is the ability to know oneself, to be self-reflective, to understand ones own feelings, moods, and motivations. **Linguistic intelligence**, according to Gardner, deals with word sense and the capacity to use words powerfully in written and spoken language (for example, as in being particularly persuasive with words). **Logical-mathematical intelligence**, sometimes called scientific thinking, involves the ability to reason abstractly, order and reorder quantitatively, and recognize significant problems and solve them through inductive and deductive thinking. Gardner identified **musical intelligence** as the ability to recognize tonal patterns and to hear and use pitch that is rhythmically arranged as musical sound. Visualizing, creating, and recreating accurate mental images of the visual world as it is was described by Gardner as **spatial intelligence**.

Reporting on how the recognition of multiple intelligences and the application of multiple-intelligence theory at the New City School in St. Louis resulted in a revised curriculum, Hoerr[2] said, "We have found that multiple intelligences is more than a theory of intelligence; it is, for us, a philosophy about education with implications for how kids learn, how teachers should teach, and how schools should operate" (p. 29).

Enthusiastic about the use of multiple intelligences in developing an integrated curriculum, Armstrong[3] said,

> At times, I almost think of Gardner as an archeologist who has discovered the Rosetta stone of learning. One can use this model to teach virtually anything, from the "schwa" sound to the rain forest and back. The master code of this learning style model is simple: for whatever you wish to teach, link your instructional objective to *words, numbers* or *logic, pictures, music, the body, social*

---

[1] Gardner, H. (1993). *Multiple intelligences: The theory in practice.* New York: Basic Books.
[2] Hoerr, T. (1994). How the New City School applies the multiple intelligences. *Educational Leadership, 52* (3), 29-33.
[3] Armstrong, T. (1994). Multiple intelligences: Seven ways to approach curriculum. *Educational Leadership, 52* (3), 26-28.

interaction, and/or *personal experience.* If you can create activities that combine these intelligences in unique ways, so much the better!...

When planning a lesson, ask the right questions! Certain questions help me look at the possibilities for involving as many intelligences as possible.

Linguistic: How can I use the spoken or written word?
Logical-Mathematical: How can I bring in numbers, calculations, logic, classifications, or critical thinking?
Spatial: How can I use visual aids, visualization, color, art, metaphor, or visual organizers?
Musical: How can I bring in music or environmental sounds, or set key points in a rhythm or melody?
Bodily-Kinesthetic: how can I involve the whole body, or hands-on experiences?
Interpersonal: How can I engage students in peer or cross-age sharing, cooperative learning, or large-group simulation?
Intrapersonal: How can I evoke personal feelings or memories, or give students choices? (pp. 26-27).

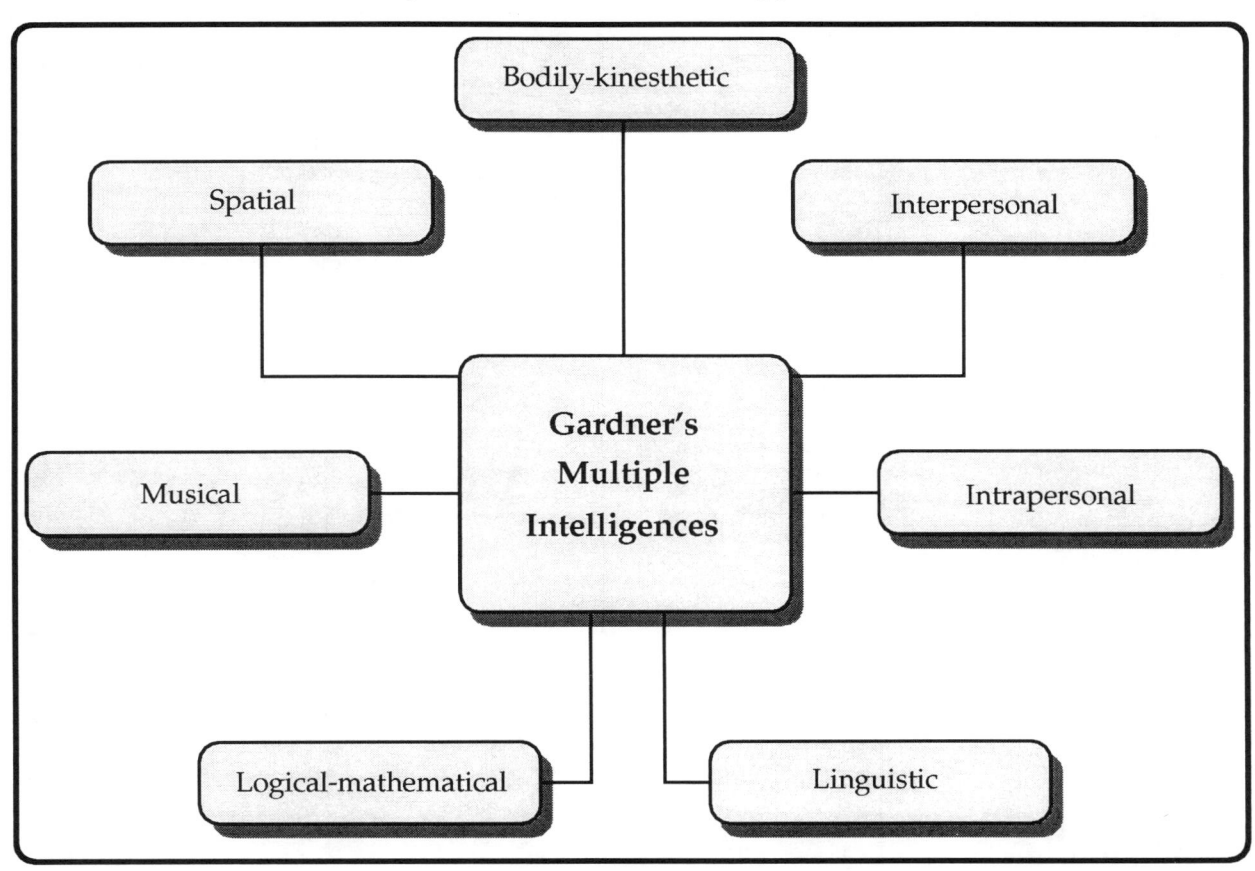

```
                    Bodily-kinesthetic
                           |
    Spatial                                Interpersonal
       \                                      /
        \                                    /
         ——  Gardner's Multiple Intelligences  ——
        /                                    \
       /                                      \
    Musical                                Intrapersonal
                           |
         Logical-mathematical        Linguistic
```

Activities for Elementary School Mathematics Page 15.1

Numerals

1	2	3	4
5	6	7	8
9	10		

Numerals Spelled Out

one

two   four

three   six

Page 15.2   Activities for Elementary School Mathematics

Numerals Spelled Out

seven	five
eight	ten
nine	zero

Numeral Pictures

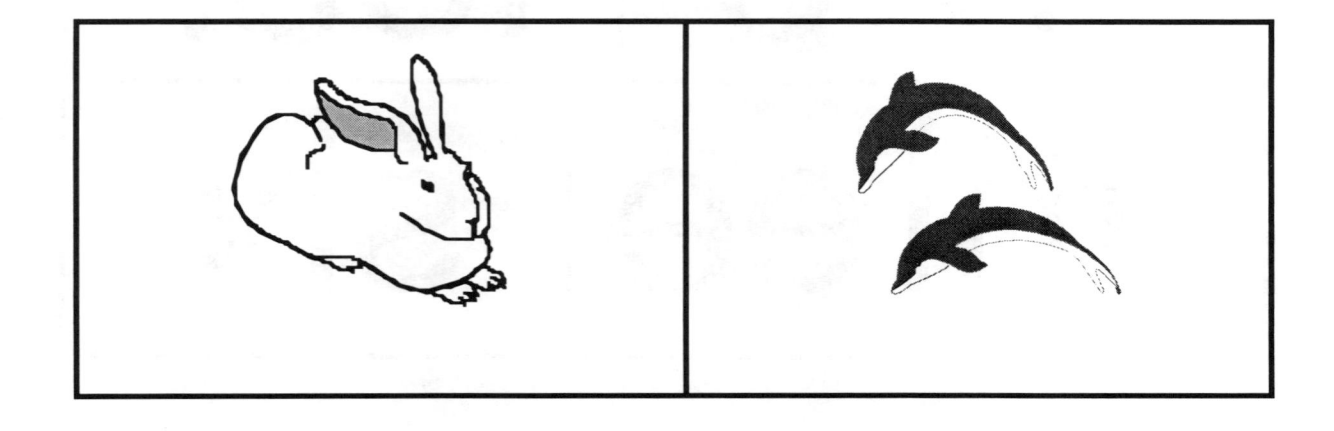

Activities for Elementary School Mathematics   Page 15.3

Numeral Pictures

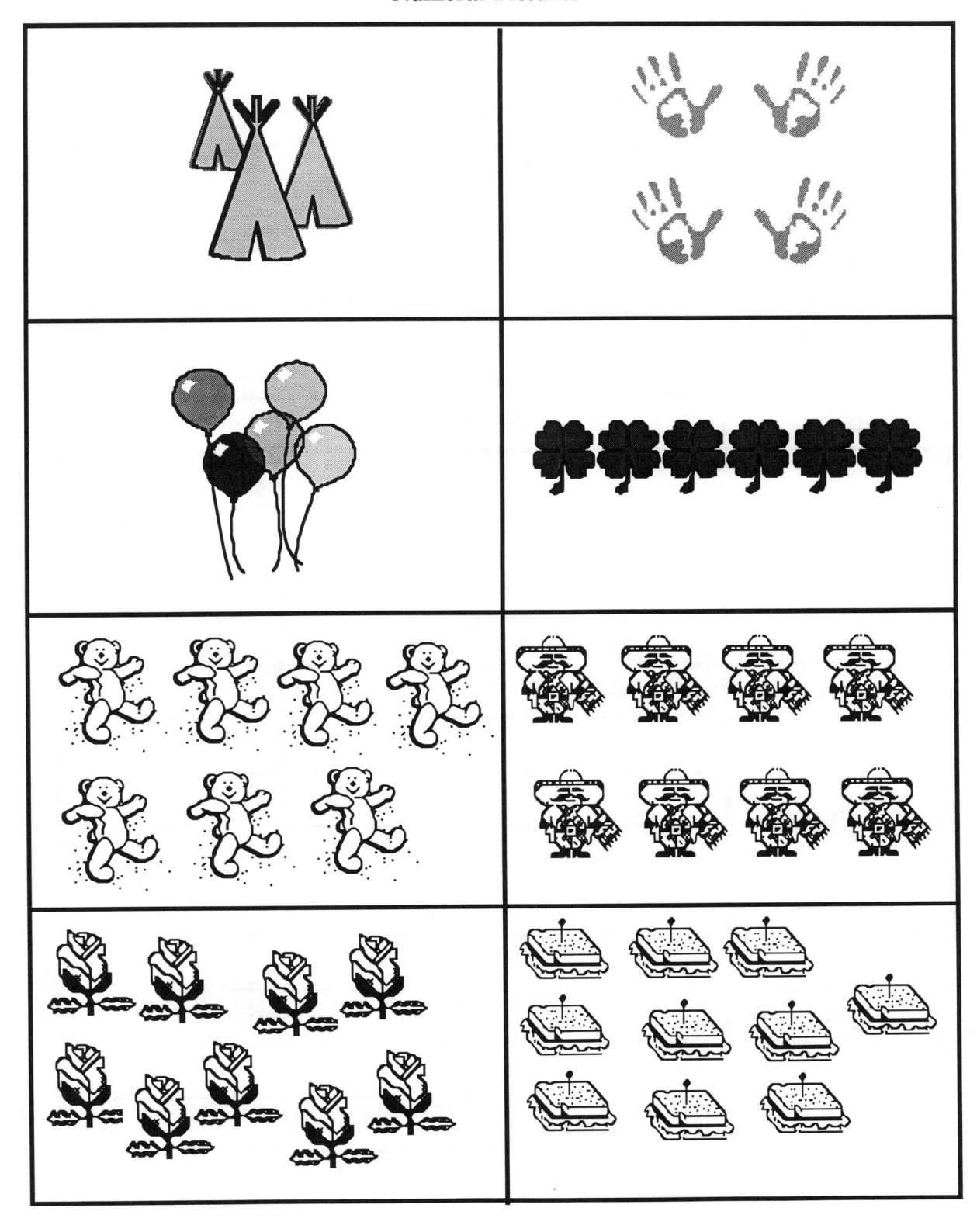

Order Form

To:	From:
_____	_____
_____	_____
_____	_____
_____	_____

Quantity	Item Description	Price
Sub Total		
Shipping	Add $1 per $12. purchase Minimum: $1.	
Sales Tax	8% on purchase price only	
TOTAL		

Activities for Elementary School Mathematics	Page 15.5

Large Numerals

1	2	3
4	5	6
7	8	9
10	11	12

Page 15.6	Activities for Elementary School Mathematics

1	2	3
4	5	6
7	8	9
10	11	12

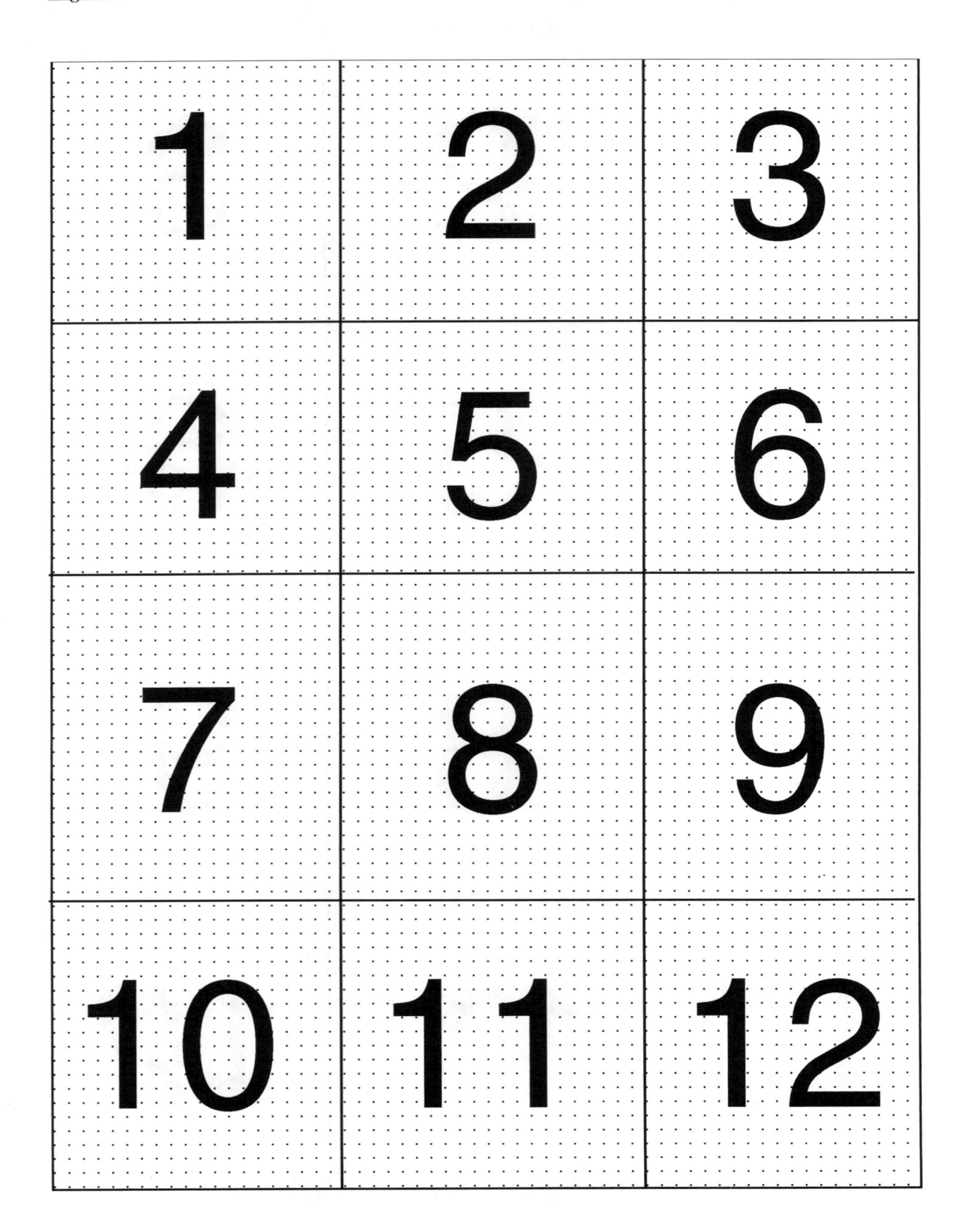

Activities for Elementary School Mathematics   Page 15.7

1	2	3
4	5	6
7	8	9
10	11	12

1	2	3
4	5	6
7	8	9
10	11	12

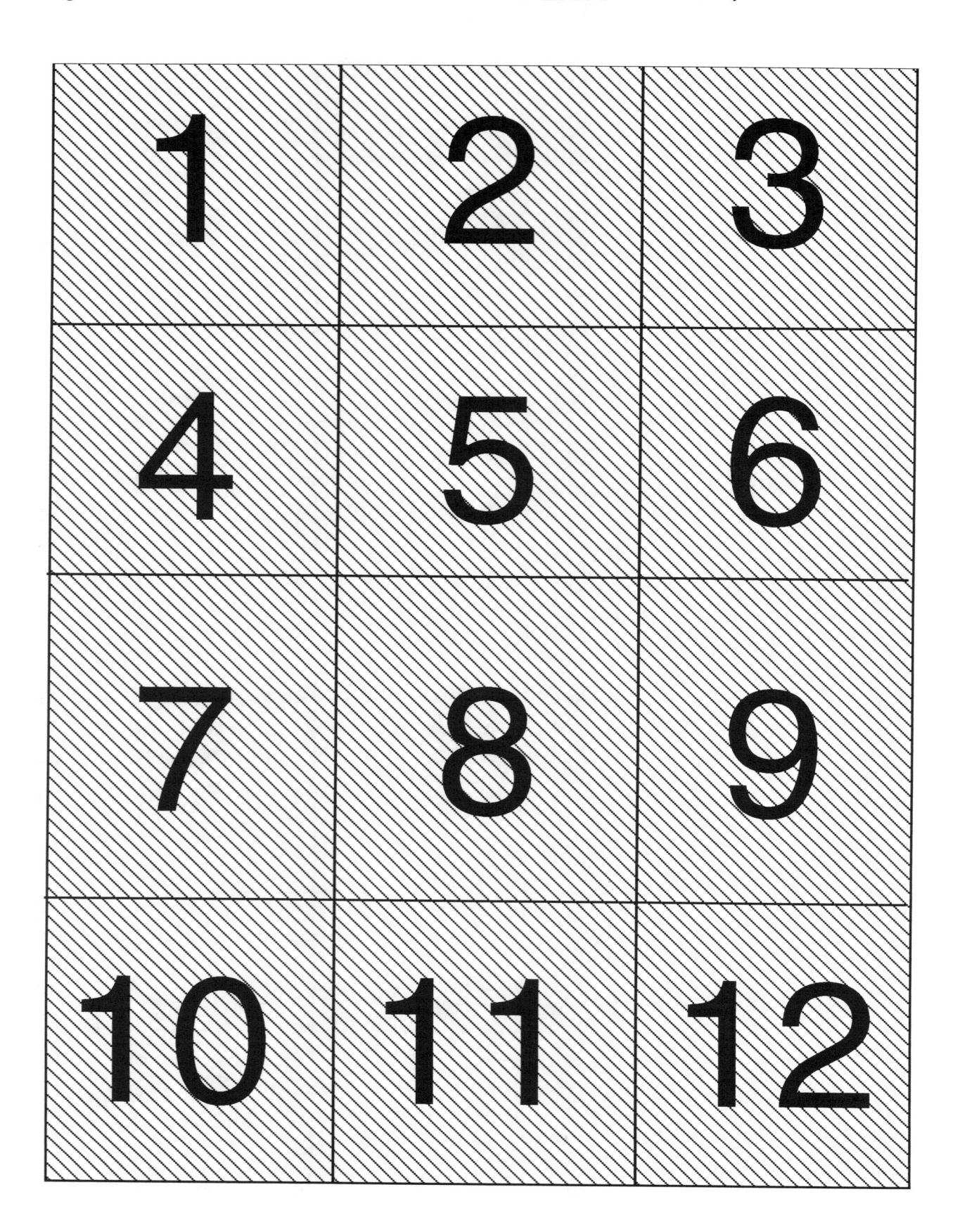

One-inch Numbered Squares

1	2	3	4	5
6	7	8	9	0

1	2	3	4	5
6	7	8	9	0

1	2	3	4	5
6	7	8	9	0

1	2	3	4	5
6	7	8	9	0

Page 15.10   Activities for Elementary School Mathematics

7 x 7 Game Board

Activities for Elementary School Mathematics   Page 15.11

Blank Bingo Card

B I N G O

Activities for Elementary School Mathematics Page 15.13

Boat Game Grid

	a	b	c	d	e	f	g	h	i	j	k	l
1												
2												
3												
4												
5												
6												
7												
8												
9												
10												

Page 15.14 Activities for Elementary School Mathematics

Geometric Shapes

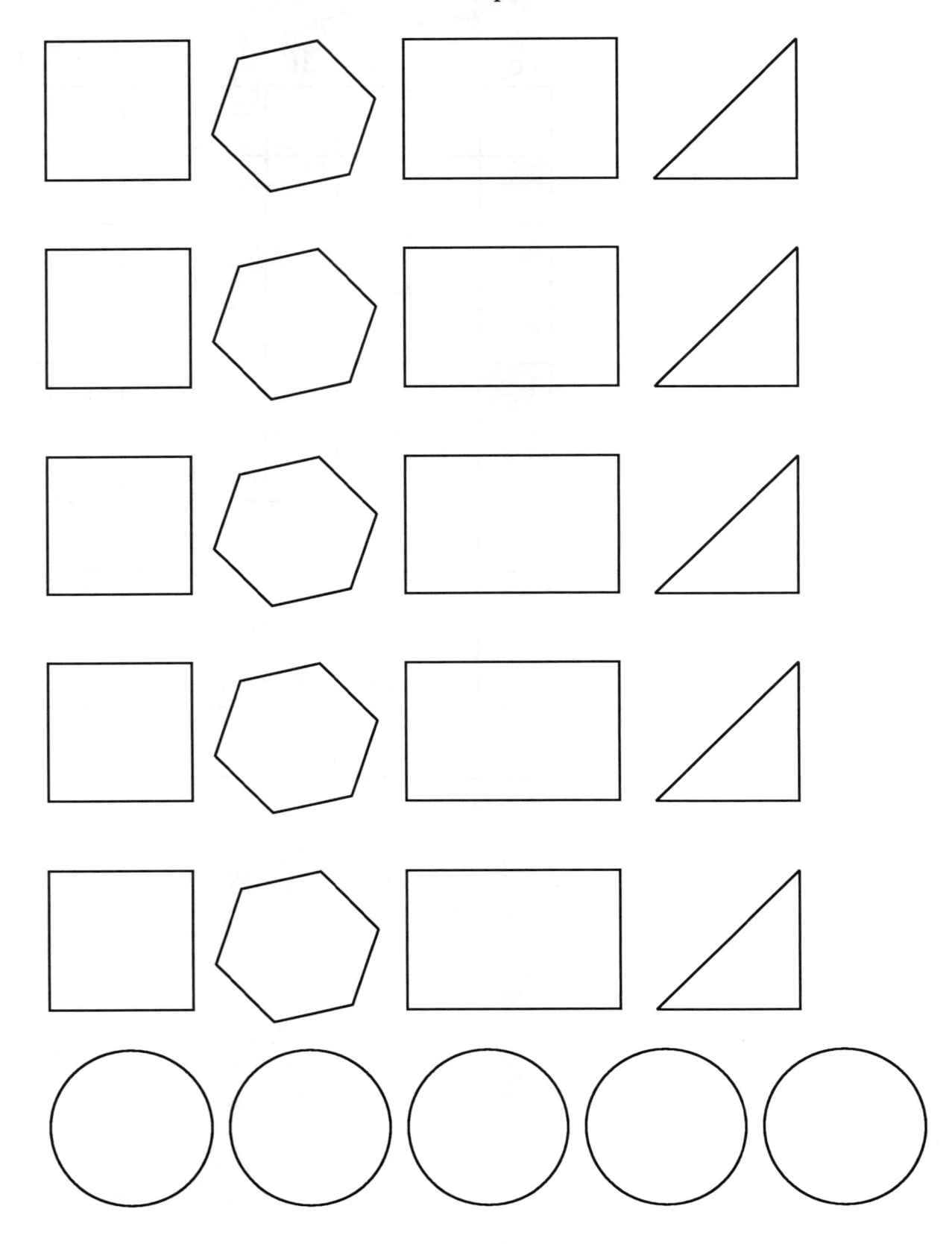

Activities for Elementary School Mathematics   Page 15.15

Triangles

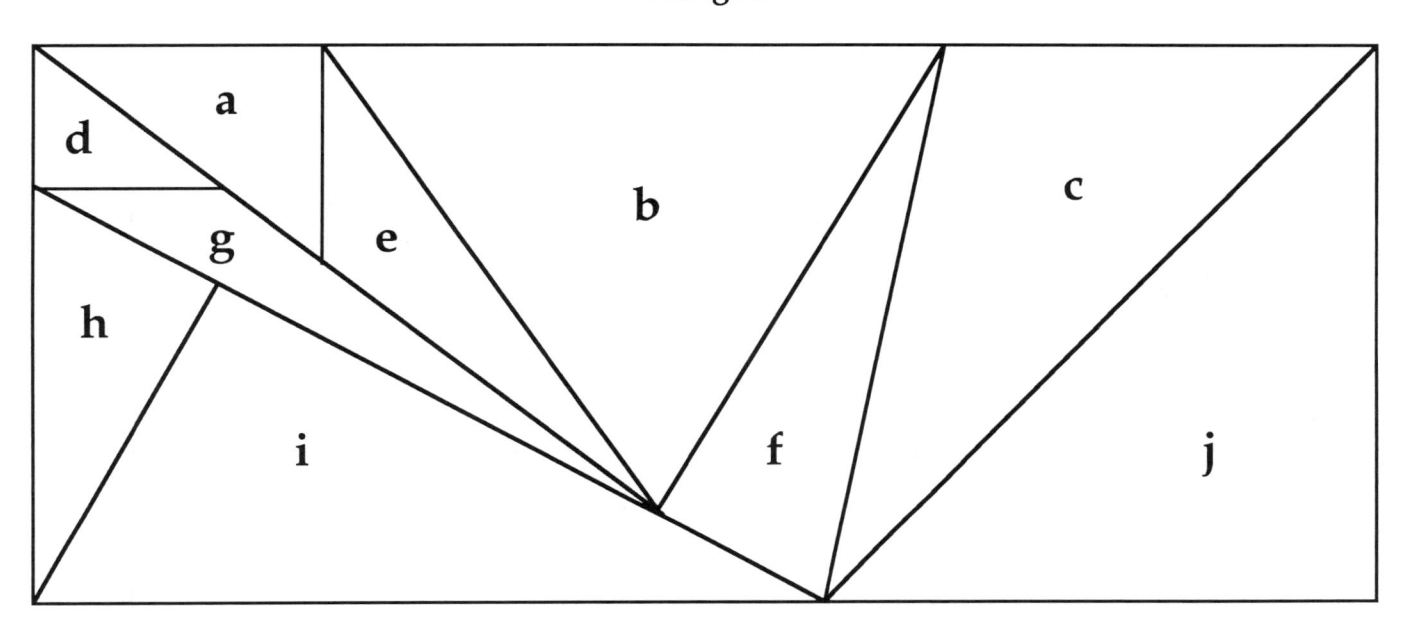

a. right isosceles triangle
b. acute equilateral triangle
c. obtuse scalene triangle
d. right isosceles triangle
e. obtuse scalene triangle
f. right scalene triangle
g. obtuse isosceles triangle
h. obtuse scalene triangle
i. acute scalene triangle
j. right isosceles triangle

Tic-tac-toe Board

Activities for Elementary School Mathematics   Page 15.17

Chart

1	2	3	4

REFERENCES

Armstrong, T. (1994). Multiple intelligences: Seven ways to approach curriculum. *Educational Leadership, 52* (3), 26-28.

Baroody, A.J., & Hume, J. (1991). Meaningful mathematics instruction: The case of fractions. *Remedial and Special Education, 12*(3), 54-68.

Brooks, J. G. (1990). Teachers and students: Constructivists forging new connections. *Educational Leadership, 48,* 68-71.

Cobb, P., Yackel, E., & Wood, T. (1992). A constructivist alternative to the representational view of mind in mathematics education. *Journal for Research in Mathematics Education, 23*(1), 2-33.

Confrey, J. (1990). What constructivism implies for teaching. In *Constructivist views on the teaching and learning of mathematics,* eds. R. B. Davis, C. A. Maher, and N. Noddings, 107-22. Reston, VA: National Council of Teachers of Mathematics.

Gallagher, J. M., & Reid, D. K. (1983). *The learning theory of Piaget and Inhelder.* Austin, TX: PRO-ED.

Gardner, H. (1993). *Multiple intelligences: The theory in practice.* New York: Basic Books.

Giordano, G. (1993). Fourth invited response: The NCTM standards: A consideration of the benefits. *Remedial and Special Education, 14*(6), 28-32.

Goldin, G. A. (1990). Epistemology, constructivism, and discovery learning in mathematics. In *Constructivist views on the teaching of mathematics,* eds. R. B. Davis et al., 31-47. Reston, VA: National Council of Teachers of Mathematics.

Hoerr, T. (1994). How the New City School applies the multiple intelligences. *Educational Leadership, 52* (3), 29-33.

Kamii, C., & DeClark, G. (1985). *Young children reinvent arithmetic: Implications of Piaget's theory.* New York: Teachers College Press.

Mathematical Sciences Education Board (MSEB) and National Research Council. (1989). *Everybody Counts: A Report to the Nation on the Future of Mathematics Education.* Washington, D.C.: National Academy Press.

National Council of Supervisors of Mathematics. (1988). *Twelve components of essential mathematics.* Minneapolis, MN: Author.

National Council of Teachers of Mathematics. (1989). *Curriculum and evaluation standards for school mathematics.* Reston, VA: Author.

National Council of Teachers of Mathematics. (1991). *Professional Standards for Teaching Mathematics.* Reston, VA: Author.

Piaget, J. (1974). *Adaptation vitale et psychologie de l'intelligence: Selection organique et phenocopie.* Paris: Hermann.

Piaget, J. (1977). *The development of thought: Equilibrium of cognitive structures.* New York: Viking Press.

Piaget, J. (1978). *Success and understanding.* Cambridge, MA: Harvard University Press.

Polya, G. (1973). *How to solve it,* 2nd ed. Princeton, NJ: Princeton University Press.

Prawat, R. S. (1992). Teachers' beliefs about teaching and learning: A constructivist perspective. *American Journal of Education, 100,* 354-395.

Pressley, M., Harris, K. R., & Marks, M.B. (1992). But good strategy instructors are constructivists! *Educational Psychology View, 4,* 3-33.

Romberg, T. A. (1993). NCTM's Standards: A rallying flag for mathematics teachers. *Educational Leadership, 50*(5), 36-41.

Vygotsky, L. S. (1978). *Mind in society: The development of higher psychological processes.* Cambridge, MA: Harvard University Press.

Wheatley, G. H. (1991). Constructivist perspectives on science and mathematics learning. *Science Education, 75*(1), 9-21.

Yackel, E., & Wheatley, G. (1990). Promoting spatial imagery in young children. *Arithmetic Teacher, 37*(6), 52-58.